Digital Marketing and Social Media Strategies for Tourism & Hospitality Organizations

Advances in Tourism Marketing

Metin Kozak, Antónia Correia and Alan Fyall

The purpose of this series of cutting-edge research-informed edited books is to introduce the reader to a range of contemporary marketing phenomena in the domain of travel and tourism. Authored by leading academics in their fields of research interest, each book will bring together a selection of related themes with individual chapters contributing theoretical, methodological, policy-related and/or practical outcomes for the reader. Each book will be introduced and brought to a conclusion by the series editors who between them have many decades of research and publishing experience. The singular aim of this advanced series of scholarly texts is to stimulate and engage readers in the fast-changing, complex and increasingly interdisciplinary nature of tourism marketing, and serve as a catalyst for future intellectual, academic, and professional-driven research agendas. This series encourages critical, participatory and humanistic approaches to research and welcomes contributions from all over the world. In particular, the series welcomes contributions from a non-Western perspective as tourism becomes truly global in both its reach and impact.

Digital Marketing and Social Media Strategies for Tourism & Hospitality Organizations

Editors:

Ahmet Bulent Ozturk and Murat Hancer

(G) Goodfellow Publishers Ltd

(G)

Published by Goodfellow Publishers Limited,
26 Home Close, Wolvercote, Oxford OX2 8PS
http://www.goodfellowpublishers.com

British Library Cataloguing in Publication Data: a catalogue record for this title is available from the British Library.

Library of Congress Catalog Card Number: on file.

ISBN: 978-1-911635-47-5

DOI: 10.23912/9781911635475-4393

Design and typesetting by P.K. McBride, www.macbride.org.uk

Cover design by Cylinder

Contents

Preface

Digital marketing and social media have revolutionized all industries by changing consumer behavior and the ways the firms operate their businesses. The tourism and hospitality industries are no exception to this transformation. Most of the today's tourism and hospitality consumers are social travelers who spend significant numbers of hours in online search for places to visit and hotels to book. Digital marketing and online social media platforms have, therefore, become the cornerstones to marketing strategy and the most viable channels for reaching potential customers and communicating with current and past customers to build an effective relationship and loyalty. In this regard, investigating the current status of digital marketing and social media utilization by both travelers and service providers, and exploring future digital marketing and social media research trends are crucially important. This book has contributions devoted to investigating important facets of digital marketing and online social media in the tourism and hospitality industries.

This book is divided into two parts. Part 1 explores the functionality of social media, social media research trends and marketing activities through social media; part 2 deals with digital marketing strategies and applications. Each chapter concentrates on a specific facet of digital marketing and/or social media by offering valuable insights for industry professionals, academicians, and students who wants to excel in important aspects of digital marketing and social media in the tourism and hospitality industries. Summary of the chapters included in this timely book are discussed below.

Chapter 1, by Seza Zerman, mainly focuses on the definition and the functions of social media, and social media usage in the hospitality and tourism industries, together with the future directions of social media use in the field.

Chapter 2, by Marwa Al-Busaidi and Khaldoon Nusair, provides a comprehensive analysis of the current state of the literature by utilizing co-word network analysis and thematic keyword analysis of 275 social media related articles in hospitality and tourism journals between 2019-2020.

Chapter 3, by Faizan Ali, Abraham Terrah, Seden Dogan and Cihan Cobanoglu, examines how social media marketing activities (SMMA) performed by the hotels may develop social identification, trust, value, and loyalty towards the hotel.

Chapter 4, by Aysegul Gunduz Songur and Gozde Turktarhan, evaluates the role of social media marketing activities in the travel planning and decision-making process of today's travelers.

Chapter 5, by Ilke Basarangil, investigateds the impacts of the Instagram effect, Bandwagon effect, and Snob effect on the transformation of tourist behavior.

Chapter 6, by Alaattin Basoda, Seden Dogan and Cihan Cobanoglu, mainly aimes to explore how emoji presence on promotional social media images differentiates the effects of travelers' reactions on their behavioral intention.

Chapter 7, by Linchi Kwok, explores content marketing in the hospitality and tourism industries, assesses the significance of content marketing, and identifies effective content marketing strategies for hospitality and tourism businesses.

Based on the Expectation-Confirmation Theory by Oliver (1980), Chapter 8, by Liza Cobos, analyzes the factors that influence customers' satisfaction with hotel branded mobile applications. It examines the contextual factors to determine the impact of mobile applications' specific characteristics on customer satisfaction with the mobile applications.

Chapter 9, by Guntekin Simsek and Sule Kaya Unal, focuses on how database marketing and big data analytics are used in the tourism industry. More specifically, it first provides the definition of database marketing and big data analysis, and then discusses their importance and benefits for tourism establishments by providing examples from the tourism industry.

Chapter 10 by Ugur Tandogan examines digitalization of customer relationship management (CRM). More specifically, this chapter evaluates CRM activities through blockchain technology in the tourism industry.

Finally, Chapter 11 by Mark Baker and Arthur Huang determines how extended reality (XR) and gamification are evaluated in the destination marketing context and identified XR and gamification research opportunities for destination marketing scholars and practitioners.

In conclusion, we would like to thank the authors who contributed to the production of this essential and timely book. We believe the chapters included in this book offer useful and important information for tourism and hospitality researchers, students, and practitioners in the context of digital marketing and social media.

Ahmet Bulent Ozturk and Murat Hancer

Author biographies

The editors

Ahmet Bulent Ozturk is an Associate Professor in the Rosen College of Hospitality Management at University of Central Florida. His research focuses on information technology adoption, digital marketing and social media, e-commerce, m-commerce, and destination marketing.

Murat Hancer is a Professor in the Rosen College of Hospitality Management at University of Central Florida. His research interests include social media marketing and management, information technology, human resources management, consumer behavior, blockchain technology and research methods.

The contributors

Seza Zerman graduated with a BS of Business Administration from Ankara University, Turkey, MSc in Hospitality and Tourism Management from Rochester Institute of Technology, USA, and PhD in Hospitality Administration from Oklahoma State University, USA. She is currently an assistant professor at Karabük University, Turkey. Her research areas are food service marketing, social media marketing, and sustainable tourism.

Marwa Al-Busaidi is a Marketing Specialist at Otaxi-Ridehailing Company. Graduated from Sultan Qaboos University with a bachelor's degree in Marketing and Information Systems.

Khaldoon Nusair is currently a Professor of Marketing and the Asst. Dean for Postgraduate Studies and Research in the College of Economics & Political Science at Sultan Qaboos University. Prof. Nusair was an Associate Professor at University of Central Florida/USA (2007-2015). He is the author of one book and more than 50 refereed papers in leading academic journals.

Faizan Ali is a marketing professor and graduate coordinator in the School of Hospitality and Tourism Management in Sarasota-Manatee. He began teaching here in 2016, coming from the Dedman School of Hospitality, Florida State University, where he served as a post-doctoral scholar. He is an active member of various hospitality associations.

Abraham Terrah is a Graduate Research Assistant at the Muma College of Business, University of South Florida. His research interests are consumer behavior, services marketing, tourism management, and hospitality technologies.

Seden Dogan is an Associate Professor in the Faculty of Tourism, at the Ondokuz Mayis University, Turkey. She worked in different 5-star hotels and tour operators between 1998-2013. She earned her Ph.D. from Akdeniz University in 2016. She is the co-author of more than 35 journal articles, more than 10 book chapters, and more than 20 conference papers.

Cihan Cobanoglu is the McKibbon Endowed Chair Professor of the School of Hospitality and Tourism Management (SHTM) in the Muma College of Business at the University of South Florida, who also serves as the Director of the M3 Center for Hospitality Technology and Innovation and Coordinator of International Programs for the SHTM.

Aysegul Gunduz Songur's research interests are within Hospitality and Tourism Marketing, Destination Marketing, Public Relations and CSR Strategies in Hospitality Industry. She is currently serving as an Adjunct Faculty teaching a variety of courses at the University of Central Florida and University of South Florida.

Gozde Turktarhan is a visiting research scholar at University of South Florida Muma College of Business M3 Center Technology and Innovation for the Hospitality and part time assistant professor at the International Final University in TRNC. Her research areas are smart destinations and tourism technologies.

İlke Basarangil received a master's degree in Tourism and Hotel Management at Mersin University Social Science Institute in 2004 and she completed her PhD at Anadolu University in 2013. Currently she is working as an Associate Professor in the Department of Tourism Guidance at Kırklareli University, Turkey. Her research interests include leisure and recreation, entertainment marketing and destination marketing.

Alaattin Basoda is an Assistant Professor in the Faculty of Tourism, University of Aksaray, Turkey. He received his Ph.D. in Tourism Management from the Social Science Institute, University of Selcuk, Turkey and a Master's in Tourism Management from Nevsehir University, Turkey. Dr. Basoda has also licenced as a tourist guide.

Linchi Kwok is an associate professor at The Collins College of Hospitality Management at Cal Poly Pomona and a visiting professor at Sun Yat-Sen University. His research won several best paper awards at academic conferences. Previously, he worked at Syracuse University, Rochester Institute of Technology, Texas Tech University, and Marriott.

Liza Cobos is an Assistant Professor of Hospitality Leadership at Missouri State University. She publishes and presents in national and international conferences in the areas of consumer-facing technology in the lodging industry. As a researcher, she seeks opportunities to combine her industry experience to pursue practical solutions to current industry challenges.

Güntekin Simsek holds an MA, PhD in Tourism Management and MSc in Computer Forensics, and is an Associate Professor at Adnan Menderes University. Şimşek, who was temporarily assigned to Kyrgyzstan-Turkey Manas University in 2019, works on tourism-based technologies, CRS/GDS, semantic web applications, electronic commerce, e-business, software quality, information security and forensics. He is a referee in scientific journals.

Sule Kaya Unal completed her master's degree in Tourism Management at Adnan Menderes University and continues PhD in Tourism Management at the same university. She works as a lecturer in the Civil Aviation Cabin Services Program at Izmir Kavram Vocational School. Her study fields are information technologies, global distribution systems, airport, ground handling and e-tourism.

V. Ugur Tandogan is an associate professor at the Tourism Faculty of Adnan Menderes University in Turkey, His research focuses on finance, cost analysis and blockchain technology in tourism.

Mark Baker is a graduate research assistant at the University of Central Florida, pursuing a Master of Science in information systems and business analytics. His research interests include business analytics, unsupervised learning in marketing, anomaly detection, and forecasting.

Arthur Huang is an assistant professor at the Rosen College of Hospitality Management and College of Engineering and Computer Science, University of Central Florida. As an interdisciplinary researcher, Dr. Huang focuses on studying artificial intelligence, data-driven decision-making, digital economy, and social well-being.

1 Definition and functionality of social media

Seza Zerman

Introduction

Online users' behaviors significantly changed after Web 2.0 was introduced, and this change forced companies to interact with their customers via new tools, called social media. After the shift from personal, static and view-only web sites to collaborative, dynamic and interactive platforms, traditional communication and sale steps are no longer acceptable (Chan & Guillet, 2011: 345). Therefore, starting from the 2000s, social media has become an essential instrument for marketing (Aydin, 2020: 1). Accordingly, some of the social media tools, such as online review sites and consumer-created content become the main information sources for the hospitality and tourism industry and its stakeholders (Moro & Rita, 2018: 344).

Unlike traditional media, where the information flows one-way from the company to individuals, social media allows interactive and interconnected communication among individuals, companies, and other stakeholders. Companies and services have been closely following the current development of social media to take the advantage of this cost-effective tool. Social media also provides a variety of opportunities for small and/or local companies, whose budgets are too limited to be a part of traditional media through advertising or professional marketing activities. From individual entrepreneurs to million-dollar companies, social media now offers a great opportunity that provides similar chances of market success, regardless of the promotion and communication budgets.

Among different social media tools, the most common and popular ones are blogs, microblogs (Twitter), social networks (LinkedIn and Facebook), media sharing domains (YouTube), review sites (Yelp, TripAdvisor), and forums (Zarrella, 2009: 3). All these platforms provide countless opportunities to companies, customers, and stakeholders for information search and marketing. Social media now provide opportunities in these platforms to communicate, create, join, collaborate, socialize, share, play, buy and sell, and learn (Tuten, 2020: 5) for both individuals and organizations.

The hospitality and tourism field is no exception. The concept of social media can be defined as an 'equalizer' (Zarrella, 2009: 7) for both small and local businesses, and national and international brands. Big companies use social media to support their professional marketing activities for integrated marketing communication; small companies have now the chance to reach out to their current or potential users, since both have similar opportunities via social media platforms. Local restaurants, destinations, and events are actively engaging with their customers and utilizing their reviews as trustable testimony.

Social media also equalizes different demographic groups as they share similar usage rates; women (68%) and men (62%); Whites (65%), Hispanics (65%) and African Americans (56%); rural (58%), suburban (68%), and urban (64%) residents utilize social media in similar numbers (Perrin, 2015: 3). This similarity among different sizes of companies and different demographic groups forms the power of social media to reach out to a variety of customers with different types of services. The power of social media also makes it attractive to learn more for both academia and industry.

This chapter of the book mainly focuses on the definition of social media, the functions and the social media usage in the hospitality and tourism industry, together with the future directions of social media use in the field.

Literature review

The influence of communication and interaction among people, via Word of Mouth (WOM), is one of the most important means of getting information. This influence is even more important if the product is intangible, and trial and evaluation is not possible before the actual consumption (Litvin *et al.*, 2008: 458). The transformation of the Internet and technology has led the shift of the Internet from a static tool for digital publishing allowing only limited one-to-many communication, to interactive collaboration platform that allows many-to-many communication (Kiráľová & Pavlíčeka, 2015: 364). The Internet also transformed the nature of WOM. The new format is a digitally accessible form of WOM, known as Electronic Word of Mouth (EWOM). Accordingly, the influence of one consumer on other customers has been transformed as well (Litvin *et al.*, 2008: 469) and has become technology-based.

As the Internet becomes more common for not only finding information or shopping but also for sharing experiences, WOM also shifted to the electronic version. While the traditional word of mouth (WOM) is defined as type of communication between customers about the products, services, or companies without commercial purposes (Litvin *et al.*, 2008: 465); electronic word of mouth (EWOM) is expressed as the form of communication among customers and businesses that is available on the Internet (Hennig-Thurau *et al.*, 2004: 39). EWOM is now the major element of social media and the most critical source of information for consumers' decision-making processes (Litvin *et al.*, 2008: 461; Živković *et al.*, 2014: 758), because while WOM is a critical way of obtaining information, the electronic version, EWOM, allows users to generate and form their own content via the Internet. This content that is available to large group of people in different versions of technology and electronic devices is known as User Generated Content (UGC) (Chu *et al.*, 2020: 3419; Doğan *et al.*, 2020: 1506). In short, today's Internet users can globally exchange their travel-related information by being a part of EWOM through User Generated Content (Chu *et al.*, 2020: 3419).

Due to the development of technology, customers are now capable of creating their own domains or sharing their content on the well-established domains. User generated content, therefore, is the way that customers add their opinion and share it with others at low or no cost (American Marketing Association, 2017). This user generated content is now shared on Internet platforms through posting, discussions, and blogging, allowing the other users or the business and companies to be a part of the process, and these platforms are called social media (American Marketing Association, 2017; Blackshaw, 2006: 179).

Easy and convenient access to connected devices such as personal computers, smart phones, and even wearable devices such as smart watches, lead customers to live in a virtuality and create and share content regardless of their location, making the Internet the *backbone* of the society (Tuten, 2020: 4). In addition, today's consumers have a large number of options to choose from and limited time to devote to make choices (Kiráľová & Pavlíčeka, 2015: 359), which increases the need for utilizing some previous experience of others. Therefore, the two groups, the first one that needs information and experience of others, and the second group that helps others by sharing their experience voluntarily, are able to meet in a platform where they can communicate, interact, and exchange information. The platform that brings these two groups together is social media. The platform extends customer engagement and involvement both in hospitality and other fields (Harrigan *et al.*, 2017: 607).

Social media and Web 2.0

To explain and define what social media is, it is essential to clearly define what social media is not. It is vital to emphasize that although the terms are used interchangeably, social media is different from Web 2.0 and User Generated Content (Kaplan & Haenlein, 2010: 60). The relationship, however, can be summarized as Web 2.0 platforms currently serve as platforms for social media services, and user generated content fuels social media. Therefore, groups and individuals create their own content by utilizing tools that are provided by social media (Obar & Wildman, 2015: 746,747). Kaplan and Haenlein (2010: 61) define user generated content as a broader term, and propose that the different ways people utilize social media form user generated content.

For individuals to clearly understand the relationship among social media, user generated content, and web technologies, the latter should be explained first. The world wide web can be viewed in three groups: Web 1.0, the web of cognition; Web 2.0, the web of human communication; and Web 3.0, the web of co-operation. The classification is not only based on the technical aspects but also based on the social dynamics and processes (Fuchs *et al.*, 2010: 43). More specifically, Web 1.0 is a network of connected information, in a static environment where web-users were passively and simply the consumers of the information. Web 2.0 provided a major shift of the platform from the network of information to the network of people and information, providing more control and freedom to the users in a more dynamic and active way. The next step, Web 3.0 will allow an interaction not only among people and groups but also between people and machines (Tuten, 2020: 11). Among these, Web 2.0 could be defined as the tool that allows social media evolution (Kaplan & Haenlein, 2010: 61).

Specifically in the tourism industry, the recent development and extensive use of Web 2.0 brings an essential use of social media in different areas (Sotiriadis, 2017: 216). Accordingly, the term Web 2.0 is revised and called *Travel 2.0*, referring to the application of travel-related technological tools (Del Chiappa, 2011: 331). Philip C. Wolf is credited as the first using the term *Travel 2.0*, which is accepted as the tailored term of Web 2.0 for the hospitality and tourism industry (Miguéns *et al.*, 2008: 2). As Web 2.0 is defined as the platform of social media evolution (Kaplan & Haenlein, 2010: 60, 67), in the hospitality and tourism industry, Travel 2.0 applications provide opportunities for both businesses and travelers to obtain information and evaluate the information and the services.

The definition of social media

One can define social media as a hybrid element of promotion efforts, which shifts the communication from company-to-customer communication to customer-to-customer communication (Mangold and Faulds, 2009: 364) or even customer-to-company communication. Therefore, the shift is not only from traditional WOM to EWOM, but also from one-way, to two-way communication. One-way communication is from company-to-customer communication, whereas two-way communication can be explained as customer-to-customer or customer-to-company communication.

Although the shift that social media brought is commonly accepted, the answer of "*what social media are and, where social media begin and end*" is not straightforward (Obar & Wildman, 2015: 745); it is challenging in contrast. There are two main reasons for this challenge: the first one is the sharp and rapid change of technology (Kaplan & Haenlein, 2010: 60; Obar & Wildman, 2015: 745). It has only been two decades since Facebook started a new part in our lives, and since then Twitter, Instagram, TikTok, and other travel and review sites have become the inevitable parts of our decision and evaluation process in just a few years. In this rapidly changing environment, it is not easy to establish the boundary of social media.

The variety of possible and accessible technologies that enable interaction among people is the second reason why social media and the boundaries of social media are not easy to define. If social media is defined as the platform that allows social interaction via technology, it raises the question of whether or not telephone and email should be considered as social media platforms (Obar & Wildman, 2015: 746). In other words, any communication through technological tools, such as telephone and email can be considered as social media interaction based on different definitions of social media. Therefore, different points of views could establish different social media definitions.

In order to clearly define social media and establish the boundaries of social media, the traditional media should be clearly defined and understood. The term media simply refers to the different ways of communication, including mass media and personal media. It can be owned (such as company webpages), paid (such as advertisements) or created (social media). Social media, therefore, goes beyond traditional media and initiates and facilitates interaction among countless of people and groups (Tuten, 2020: 5). Social media use is a specific way of utilizing digital media and the Internet, and it is significantly different from conventional media usage through the process that social media users utilize the Internet to socialize, connect, communicate and interact (Correa et al., 2010: 248).

Thus, social media is defined as '*culture of connectivity*' by Van Dijck (2013), meaning that it is the new way to create a connection not only business to businesses, but also business to consumer, and consumers to consumers, and even consumers to machines (Tuten, 2020: 9). For companies, the social media is the profitable way of utilizing the technology (Kaplan & Haenlein, 2010: 59). For individuals, through its technological efficiency and mobility, social media provides an online tool for interdependent and interconnected people and communities to communicate and collaborate (Tuten, 2020: 9).

"*Social media is a group of Internet-based applications that build on the ideological and technological foundations of Web 2.0, and that allow the creation and exchange of user generated content.*" (Kaplan & Haenlein, 2010: 61). The social interaction takes place through social channels and related vehicles, called zones of social media (Tuten, 2020: 12). Although there is no commonly agreed definition of social media, it is logical to propose that blogs, social networking sites, and the sites for sharing feedback form the main categories of social media (Sotiriadis, 2017: 182).

The dissimilarities between traditional media and social media also separate and group the users of social media. People who grew up a couple of decades ago are more familiar with traditional media compared to social media. As this group is unfamiliar with technology such as wearable devices and new forms of communication such as video calls, they can be considered as *guests* in today's technological lifestyle. Social media users can be either the group that grew up with the Internet or had to adapt to it later in their lifetime. In the notable work "Digital natives, digital immigrants", Prensky (2001: 1) classified users of the Internet in two groups. The first, *digital immigrants*, encountered the Internet later in their life, which forced them to explore, learn it from the manual and apply the knowledge. The second group, *digital natives*, on the other hand, grew up with the Internet; therefore, digital way is their native language (Prensky, 2001: 1). Kaplan and Haenlein (2010: 61) propose that the rise of the digital natives' generation, is the social driver of user generated content because this group has considerable knowledge, and they are eager for online interaction. Whether digital natives or digital immigrants, social media works only when people are engaged by participating, creating and sharing content (Tuten, 2020: 4). Although digital natives' attention span is limited for many traditional ways of teaching/ learning, their span is way longer for games and interactive activities (Prensky, 2001: 5). At this point, social media is the way of interactivity and connection through its various application for various groups.

In short, although the definition and the boundaries of social media are not clear, social media can be defined as an online platform for individuals, profit-making and non-profit organizations, and various stakeholders to communicate and collaborate in interrelated and interconnected ways using the development of technological devices and generations of the Web. Different types of social media and different applications in various platforms serve different purposes in particular domains. However, the main functions of social media remain the same, and these functions are building a connection, creating an interaction, and exchanging information among stakeholders, and providing insight about products and services.

Types of social media

As stated above, to define and classify social media is not easy due to the rapid chance of technology. In addition, similarities, differences, and the boundaries of the terms among social media, user generated content, Web 2.0, or Travel 2.0 are still blurry and these terms are generally used interchangeably. For example, the content added by users on the user generated sites, such as personal information, photographs, videos, likes, comments and so on are fueling social media sites (Obar & Wildman, 2015: 745).

In addition, different use of social media introduced variations of social media tools. For example, recently, social media has been utilized by organizations and companies in order to facilitate formal communication among employees, and this tool is called *enterprise social media (ESM).* Unlike the social media communication between companies and customers, the internal communication between company and its employees, ESM, is the transformational way to facilitate the communication among employees via various channels such as emails and QandA forums (Leonardi *et al.*, 2013: 3). Furthermore, recent development of hi-tech devices also introduced new social media types. For example, the launch of iPhone in 2007 exploded the mobile marketing activities. Accordingly, new form of mobile marketing is associated with social media and that is called *mobile social media*. This form of social media allows companies to track and monitor customers and market their service and product accordingly (Seth, 2012: 10). In short, the addition of new concepts, such as enterprise social media and mobile social media, is a dynamic and inevitable aspect of social media definition and function.

Social media is a part of a larger media group where it can be classified under three main categories: bought media (e.g., advertisement); owned media (e.g., web pages of brands and/or products); and earned media (e.g., social media) (Kietzmann *et al.*, 2011: 249). As the importance of social media is increasing, it is further classified under two main groups as owned social media (OSM) and earned social media (ESM). The businesses and companies are the main actors that create and control OSM, with limited control of the users and customers. The examples are blogs, YouTube channels and fan pages. ESM is also related with the businesses and brands, but unlike OSM, ESM gives control and manipulation to the users and customers rather than brand owners (Colicev *et al.*, 2018: 39). Travel sites such as TripAdvisor and Yelp are popular examples of the latter social media type. Okazaki *et al*. (2017: 111, 112) also propose another classification for social media grouping: it can be either personal social media, such as a Facebook page for a small circle of individuals, or public social media that is available for a larger group of users, as experience sharing sites such as TripAdvisor.

One comprehensive grouping for social media was proposed by Van Dijck (2013: 8), under four main categories:

- Social network sites, for example Facebook,
- User-generated content, for example YouTube,
- Marketing and trading sites, for example Amazon, and
- Play and game sites, for example Angry Birds.

Tuten (2020: 17) groups social media under four main categories:

- Social community (Twitter, Facebook, LinkedIn, Instagram),
- Social publishing (blogs and media sites, e.g. YouTube and SlideShare),
- Social commerce (Facebook, Etsy, TripAdvisor),
- Social entertainment (Spotify, YouTube),

Another grouping is proposed by Chan and Guillet (2011: 347, 348):

- Blogs and micro-blogs,
- Social network sites,
- Virtual worlds,

- Collaborative projects,
- Content community sites,
- Feedback sites.

As it can be seen from the classifications above, there is no certain way to group social media applications. Furthermore, the rapid evolution and development of social media would require new classification and grouping in years to come (Kaplan & Haenlein, 2010: 61, 62). Although classifications are different, for various purposes the majority of the social media serves one united purpose, that is social media marketing. Tuten (2020: 19) defines social media marketing as the way that the organizations and companies utilize the components of social media, technologies, tools, and software, and how they establish, communicate, distribute, and exchange any type of offering that has a potential value. Unlike traditional marketing, where attention is obtained via orientation on mass media channels; social media marketing targets interaction, participation, dialogue, and sharing to gain attention on niche and social media channels (Tuten, 2020: 19).

It is important to clearly understand that, although the social media is a strong and effective domain for marketing, some risks are associated with social media, such as it being time-consuming and overwhelming (Akehurst, 2009: 51); therefore, it should be carefully designed and monitored. Some possible solutions being used in social media marketing are data mining, sentiment analysis, data visualization and storytelling.

In channels such as social media, consumers do not passively listen to the brand story, but actively tell the brand story. In other words, consumers are authors of the brand stories in social media environment (Gensler *et al.*, 2013: 244). In the hospitality industry, thanks to social media, travelers now tell the stories in each stage of their experience, from researching the trip, to illustrating and sharing their experience (Leung *et al.*, 2013: 5). Accordingly, destinations become closer for any potential visitors, in any part of the world, through the reviews, photographs, and stories that the visitors created (Kiráľová & Pavlíčeka, 2015: 359).

Although storytelling in social media can help to design the brand and its components, such as communication strategies and promotion tools, the strategies should be carefully monitored and measured. Considerations such as social media metrics and big data analysis are the topics to be evaluated for the users to operate social media effectively (Minazzi, 2015: 137) because in addition to its connection function, social media is also used to create a database for data mining (Van Dijck, 2013: 4).

Advancements in technology and the Internet changed the traditional rules of marketing and brought great advantages to many industries, with hospitality and tourism being one of the most affected and benefited industries (Litvin *et al.*, 2005: 156). Now, online marketing strategies should start with an understanding of social media because customers find social media information more trustworthy compared to the company's information (Castronovo & Huang, 2012: 120). The online presentation of hospitality products or services is a more dynamic social construction as it is accumulative and generative, owing to a company, customers, and their interactions (Hunter, 2016: 222). The rapid growth in social media has made user generated content a virtual currency to manage reputation and competition for the hospitality industry (Ma *et al.*, 2018: 3303). The application of social media content and tools has gained importance in hospitality and tourism, thus requiring further understanding of the social media concept in this particular field.

Social media in the hospitality industry

The evolution of social media in the hospitality and tourism area is exponential (Mehraliyev *et al.*, 2020: 215), because social media is an essential tool and has implications such as holistic and real time insight, and credible information for the hospitality and tourism industry. Social media *per se* provides a cost-effective platform for hospitality companies for interaction and engagement, without the requirement for any additional hardware or software (Seth, 2012: 9).

One role of social media is offering unfiltered data, provided by different travelers from all around the world (Chan & Guillet, 2011: 345, 346). WOM communication has already been utilized in the hospitality industry, where the products and services cannot be tried before the actual experiences. The online form of WOM, EWOM through social media, is fundamental for the hospitality and tourism industry for several reasons: first, hospitality products are intangible, and it is not easy to evaluate the service prior to the actual consumption; second, the service and products of the industry are generally high involvement products and the purchase of such products is not routine and requires an intense process (Del Chiappa, 2011: 332). These characteristics lead travelers to seek credible and trustable information via various channels, either traditional WOM or EWOM through social media and travel 2.0 applications.

Among many forms of social media tools, the main forms are blogging sites, micro blogging sites, social news sites, media sharing sites, forums, sites that create virtual worlds, and rating and review sites (Zarrella, 2009: 3). The last group, rating and review sites include Yelp, TripAdvisor and other sites, in which the users and customers share their experience as independent agents. These independent agents, current or potential customers, utilize social media in interactive and independent ways, such as locating the best restaurant in a city based on the previous reviews, and keeping a travel diary with photos, videos, and ratings (Tuten, 2020: 14).

Social media is beneficial from the two sides of the hospitality industry. For consumers, it is an essential source of information and decision making based on the information, and from the companies' side it is beneficial for tourism promotion and interaction with customers as a tourism management and marketing strategy (Zeng & Gerritsen, 2014: 28). In short, social media is mutually beneficial for the two parties of the industry: it is a way for tourists to express their needs, and for service providers to monitor the needs and feedback of customers (Sotiriadis, 2017: 179, 180). Accordingly, the research in the hospitality domain reveals two main categories in sharing the experience in social media: namely the consumer's and the provider's perspectives (Sotiriadis, 2017: 185). The former explains the antecedents of information-sharing behavior and the impact of the information on other tourists, where the latter explains the impact of the information on business for marketing and management functions.

Customers utilize social media in the planning phase, travel stage, and post-trip stage, whereas the suppliers in the field utilize it for promoting their services, managing their business, searching the activities and potential businesses, communicating with stakeholders, and distributing the service and product effectively (Leung *et al.*, 2013: 6, 10). In a competitive hospitality and tourism industry it is vital for companies to utilize social media for more engaged stakeholders (Harrigan *et al.*, 2017: 600). The distinction between consumers and service providers gets blurry, and social media becomes a common and collaborative platform for customer to customer, company to customer, and customer to company communication.

It is commonly agreed that travelers utilize social media for generating travel ideas and plans, making the actual plans, and sharing post travel experiences, and social media can facilitate for those activities and even alter the choices of the travelers after the interaction on social media (Del Chiappa, 2011: 332). Today's tourists actively engaged with technology for their decision-making process via a comprehensive Web 2.0 structure; around 50% of travelers already have variety of travel applications to search information about destination before their actual trip (Živković *et al.*, 2014: 758).

Based on a comprehensive literature review, Zeng and Gerritsen (2014: 32, 33) propose that social media is an essential tool to obtain in-depth knowledge in various areas, and the authors classify social media research in the hospitality industry under the following groups:

- Tourist behavior,
- Marketing and destination management,
- Tourism culture and knowledge,
- Crisis management,
- User generated content,
- Issues in social media usage.

Social media is also used in different industries and settings such as the hotel industry, the tourism experience, destinations, restaurants and dining, the travel trade, airlines, and events and festivals. It is fundamental to clarify the use and adaptation of social media in these different settings for several reasons: first, social media has an impact on tourist behavior and perception, and on management and marketing initiatives of service providers (Sotiriadis, 2017: 180); and second, it can provide tremendous opportunities for small businesses in the industry at no or low cost. Therefore, understanding the areas that the hospitality industry can utilize and focus on becomes the main consideration for academics and industry professionals.

Brand management and social media

The traditional approach to brand management explains consumer behavior from the information processing theories perspective and defines the brand as an asset that is created and controlled by companies, assuming every single consumer will understand and evaluate the brand in the way that the company established. On the other hand, the post-modern view, that is built on the consumer culture theory, proposes that consumers are not passive absorbers, in fact, they are co-creators of the meaning of the brand (Gensler *et al.*, 2013: 243). Thus, the shift from read-only content, that is Web 1.0, to collaborative content, which is Web 2.0, explains not only the technological evolution of Web but also the evolution of brand management. However, although the relationship between social media and brand management is well-established topic, how social media can be utilized, and how it can be incorporated with brand management and branding strategies are still less known topics, which should be researched more in-depth (Moro & Rita, 2018: 358). In the hospitality and tourism domain, companies started to utilize social media either to form a brand identity through social media, or to position social media as a part of their integrated marketing communication and brand management activities. Especially, for intangible services, where the tangible components are not available to evaluate the service, social media can establish and sustain the brand image in the long term.

Hotel management and social media

Moro and Rita (2018: 345) argue that accommodation is the most common aspect of hospitality industry in social media. From small to big, independent to chain, hotels are aware of the fact that they are dealing with new age customers (Seth, 2012: 11). Accordingly, one of the most common applications of social media in hospitality and tourism industry is marketing activities in the hotel industry. A study by Chan and Guillet, (2011: 355, 357, 358) revealed that hotels utilized social media for various stages of their marketing activities. These are attracting customers by advertising promotions and discounted services; engaging through interaction and discussion groups in virtual communities; retaining, learning, and relating. However, some major barriers, such as being unable to interact with customers, lack of commitment to social media activities, failure to provide responses to guests, and inaccurate content, prevent companies from being effective in social media (Chan & Guillet, 2011: 359). Together with technological advancement, such as online check-in, social media offers great potential to manage accommodation preferences and experience from the supply and demand points of views.

Destination management and social media

The hospitality and tourism industry is a highly competitive environment and online influences provide competitive advantages for the users who can utilize the platforms in the early stages (Litvin *et al.*, 2008: 458, 459). Strategies supported by social media and its functions can help destination gain competitive advantage (Kiráľová & Pavlíčeka, 2015: 359). Before leaving for the destination, travelers are likely to obtain information about the destination through Web 2.0 applications (Živković *et al.*, 2014: 759). This information is mainly created by the previous visitors and today's social media users collaborate to form the image of a destination (Miguéns *et al.*,

2008: 1). Therefore, social media now provides information for certain destinations and offers opportunities to compare the destinations, not only in close regions but also in different countries and continents.

Customer relationship management and social media

Despite the undeniable power of social media for branding and marketing, a limited number of businesses seem to utilize the opportunity to interact with customers. For example, less than 8% of small restaurants post content on Facebook every week by themselves, and only 2% of them advertise with Facebook ads (Eling, 2017). In addition to insufficient use of social media ads, service providers also fail to properly analyze the data and information, which is voluntarily provided by the customers. Every single interaction between users and the Internet, in other words, "the traces the users left behind," by the customers becomes a possible domain in which to capture, store, and analyze the data (Yang *et al.*, 2015). Analyzing the data on social media is a way of decoding the hospitality meaning (Ting *et al.*, 2013). Social media is the medium to gain in-depth understanding of the needs, satisfaction and service failure perception of customers.

In short,the hospitality industry is one of the biggest beneficiaries of social media, in various areas, such as brand management, hotel management, destination management, and customer relationship management. Accordingly, companies still need further understanding on social media to reach the full potential.

Conclusion

The platform that consists of different forms of social media changed both supply and demand aspects in hospitality and tourism industry. In simple terms, social media is the electronic way to create, share, analyze, interact, and communicate via the Internet, for prospective, returning, and loyal customers, as well as businesses and other stakeholders of the industry.

Social media is an essential way for people not only to search information but also to communicate and interact with other people and to be heard. The Web 2.0 technologies, referred as Travel 2.0 for the hospitality and tourism industry, social media allows people to switch from being passive consumers to being active participants. Considering the high involvement characteristics of the tourism product, the interaction among current and potential customers is essential not only for the customers and travelers but also for the other stakeholders in the industry such as businesses and policy makers. From the businesses' point of view, it provides strategic advantage for activities, performance, and competitiveness. For this reason, it is essential to define and clarify what social media is, how it can be utilized, and how to implement the current tools effectively (Sotiriadis, 2017: 216).

The speed of technological change and other competitive technologies make the definition and classification of social media complicated and challenging. However, what we exactly know is that social media is the platform that allows any form of information, satisfaction or failure, to spread within seconds to the crowd in an uncontrolled way.

Despite the power and potential of social media, it should also be clearly explained that, although social media and search engines seem to be the very first information source, the companies' web sites are still one of the sources for travelers (Verma *et al.*, 2012: 184). The major points to focus on social media can be considered from two angles: first, the points of using social media; and the second, points of being social (Kaplan & Haenlein, 2010: 65). The companies should still use their electronic sources, such as their own websites, to communicate with their customer as the first and the most official source of information.

Social media is beneficial to find information and share experiences. Accordingly, it is utilized both by the tourists and service providers. Živković *et al.* (2014: 758) name the individuals who search and share information on social media as 'participants'; the term includes both tourists and stakeholders. Thus, the future of social media use in hospitality and tourism will more likely allow interaction not only between service providers and tourists but also among all stakeholders such as local communities, policy makers, trend setters and even machines. Considering the fact that the next generation of the Web, Web 3.0, allows the interaction and collaboration between machines and people, the future of social media in the hospitality and tourism field will evaluate around Web 3.0 technologies by counting machines as stakeholders of hospitality products on social media.

Social media exist in a complicated and rapidly growing environment; people who join social media also create and contribute it (Tuten, 2020: 4). In the hospitality and tourism industry, various shareholders, including local communities, can leverage the social media for economic progress, including community engagement and local communities through social media (Zeng & Gerritsen, 2014: 34). After the development of Web 1.0 and Web 2.0, the next generation, Web 3.0, will bring the collaboration of people and machines. In other words, machines will be involved in the process of creation, collaborating, and sharing (Tuten, 2020: 10, 11). Thus, the context of user generated content, and the mining and processing of big data will eventually be the focus of the research in Web 3.0 stage (Tuten, 2020: 9). The hospitality and tourism industry, therefore, should be proactive about the Web 3.0 technologies by integrating machine interaction in different social media applications.

Although the current studies and research show that the information on social media has been offering tremendous opportunities for small businesses, big players and individuals, there are some other details that social media should be evaluated with. While it is known that personality traits are the predictors of social media usage, gender and age also affect social media behavior (Correa *et al.*, 2010: 247, 252). The credibility and trustworthiness of social media content is one of the concerns as well as legal and ethical issues, especially in tourism and hospitality context (Zeng & Gerritsen, 2014: 34). The rapid spread of technology brings some legal issues as well. These issues include but are not limited to privacy, intellectual property of the content, protection of certain groups such as children and minorities (Obar & Wildman, 2015: 10). Thus, the hospitality industry should take these issues into consideration within legal and regulatory frameworks. As social media has a power to influence and even alter the choice of the travelers, different level of credibility and trustworthiness of different Web 2.0 applications become more important to investigate (Del Chiappa, 2011: 331). Therefore, the industry should take the characteristics of social media users together with the legal aspects of it into consideration to utilize social media efficiently and gain maximum advantages from available information on social media.

In conclusion, social media, which enables companies to interact with consumers and follow the competitors, and helps customers to find, create and share information, requires professional management. Individuals, businesses, non-profit and government organizations now are on social media via different channels and for different needs. To be able to accommodate the need, new jobs are emerging in social media (Tuten, 2020: 30, 31). Considering the fact that social media data become essential to monitor, manage, and analyze, future jobs in the hospitality and tourism area, and the marketing field in general, would be more important to monitor and evaluate.

Discussion questions

1. What are the major differences between traditional media and social media? In which aspects do social media differ from traditional media?
2. What are the main characteristics of Web 1.0, Web 2.0, and Web 3.0 and what is the relationship between social media and these three web technologies?
3. What are the major areas that utilize social media in the hospitality industry? What would be the other potential areas in which social media can be utilized?
4. What are the main categories of social media? Based on your own understanding, which of these categories offers more opportunities and provides more information for the industry professionals?
5. What are the reasons for using social media? Provide two reasons for both business owners and tourists.

References

Akehurst, G. (2009) User generated content: the use of blogs for tourism organizations and tourism consumers, *Service Business*, 31 (1), 51-61.

American Marketing Association (2017) *Power to the people: Why user-generated content is so effective in marketing*, American Marketing Association. www.ama.org/partners/content/Pages/power-to-the-people-why-ugc-marketing-is-so-effective.aspx [Accessed: 9 October 2020].

Aydin, G. (2020) Social media engagement and organic post effectiveness: A roadmap for increasing the effectiveness of social media use in hospitality industry, *Journal of Hospitality Marketing & Management*, 29 (1), 1-21.

Blackshaw, P. (2006) The consumer-controlled surveillance culture https://www.clickz.com/the-consumer-controlled-surveillance-culture/69332/ [Accessed: 13 April 2019]

Castronovo, C. & Huang, L. (2012) 'Social media in an alternative marketing communication model', *Journal of Marketing Development and Competitiveness*, 6 (1), 117-134.

Chan, N.L. & Guillet, B.D. (2011) Investigation of social media marketing: How does the hotel industry in Hong Kong perform in marketing on social media websites?, *Journal of Travel & Tourism Marketing*, 28 (4), 345-368.

Chu, S.C., Deng, T. & Cheng, H. (2020) The role of social media advertising in hospitality, tourism and travel: A literature review and research agenda, *International Journal of Contemporary Hospitality Management*, 32 (11), 3419-3438.

Colicev, A., Malshe, A., Pauwels, K. & O'Connor, P. (2018) Improving consumer mindset metrics and shareholder value through social media: The different roles of owned and earned media, *Journal of Marketing*, 82 (1), 37-56.

Correa, T., Hinsley, A.W. & De Zuniga, H.G. (2010) Who interacts on the web?: The intersection of users' personality and social media use, *Computers in Human Behavior*, 26 (2), 247-253.

Del Chiappa, G. (2011) Trustworthiness of travel 2.0 applications and their influence on tourist behaviour: An empirical investigation in Italy' *ENTER*, 331-342.

Doğan, S., Basaran, M.A. & Kantarci, K. (2020) Determination of attributes affecting price-performance using fuzzy rule-based systems: Online ratings of hotels by Travel 2.0 users. *Journal of Hospitality and Tourism Technology*, **11** (2), 291-311.

Eling, J. (2017) *Restaurant trends for 2017*, Marketing4Restaurants. https://marketing4restaurants.com/33-restaurant-trends-for-2017/ [Accessed 3 January 2021].

Fuchs, C., Hofkirchner, W., Schafranek, M., Raffl, C., Sandoval, M. & Bichler, R. (2010) Theoretical foundations of the web: Cognition, communication, and co-operation. Towards an understanding of Web 1.0, 2.0, 3.0, *Future Internet*, **2** (1), 41-59.

Gensler, S., Völckner, F., Liu-Thompkins, Y. & Wiertz, C. (2013) Managing brands in the social media environment, *Journal of Interactive Marketing*, **27** (4), 242-256.

Harrigan, P., Evers, U., Miles, M. & Daly, T. (2017) Customer engagement with tourism social media brands, *Tourism Management*, **59**, 597-609.

Hennig-Thurau, T., Gwinner, K.P., Walsh, G. & Gremler, D.D. (2004) Electronic word-of-mouth via consumer-opinion platforms: What motivates consumers to articulate themselves on the internet?, *Journal of Interactive Marketing*, **18** (1), 38-52.

Hunter, W.C. (2016) The social construction of tourism online destination image: A comparative semiotic analysis of the visual representation of Seoul, *Tourism Management*, **54**, 221-229.

Kaplan, A.M. & Haenlein, M. (2010) Users of the world, unite! The challenges and opportunities of social media, *Business Horizons*, **53** (1), 59-68.

Kietzmann, J.H., Hermkens, K., McCarthy, I.P. & Silvestre, B.S. (2011) Social media? Get serious! Understanding the functional building blocks of social media, *Business Horizons*, **54** (3), 241-251.

Kiráľová, A. & Pavlíčeka, A. (2015) Development of social media strategies in tourism destination, *Procedia-Social and Behavioral Sciences*, **175**, 358-366.

Leonardi, P.M., Huysman, M. & Steinfield, C. (2013) Enterprise social media: Definition, history, and prospects for the study of social technologies in organizations, *Journal of Computer-Mediated Communication*, **19** (1), 1-19.

Leung, D., Law, R., Van Hoof, H. & Buhalis, D. (2013) Social media in tourism and hospitality: A literature review. *Journal of Travel & Tourism Marketing*, **30** (1-2), 3-22.

Litvin, S.W., Blose, J.E. & Laird, S.T. (2005) Tourists' use of restaurant webpages: Is the internet a critical marketing tool?, *Journal of Vacation Marketing*, **11** (2), 155-161.

Litvin, S.W., Goldsmith, R.E. & Pan, B. (2008) Electronic word-of-mouth in hospitality and tourism management, *Tourism Management*, **29** (3), 458-468.

Ma, E., Cheng, M. & Hsiao, A. (2018) Sentiment analysis–a review and agenda for future research in hospitality contexts, *International Journal of Contemporary Hospitality Management*, **30** (11), 3287-3308.

Mangold, W.G. & Faulds, D.J. (2009) Social media: The new hybrid element of the promotion mix, *Business Horizons*, **52** (4), 357-365.

Mehraliyev, F., Choi, Y. & King, B. (2020) Theoretical foundations of social media power in hospitality and tourism: A hierarchical model, *Cornell Hospitality Quarterly*, **62** (2), 215-231.

Miguéns, J., Baggio, R. & Costa, C. (2008) Social media and tourism destinations: TripAdvisor case study. *Advances in Tourism Research*, **26** (28), 1-6.

Minazzi, R. (2015) *Social Media Marketing in Tourism and Hospitality*. Springer International Publishing.

Moro, S. & Rita, P. (2018) Brand strategies in social media in hospitality and tourism, *International Journal of Contemporary Hospitality Management*, **20** (1), 343-364.

Obar, J.A. & Wildman, S.S. (2015) Social media definition and the governance challenge: An introduction to the special issue, *Telecommunications Policy*, **39** (9), 745-750.

Okazaki, S., Andreu, L. & Campo, S. (2017) Knowledge sharing among tourists via social media: A comparison between Facebook and TripAdvisor, *International Journal of Tourism Research*, **19** (1), 107-119.

Perrin, A. (2015) Social media usage, *Pew Research Center*, **125**, 52-68.

Prensky, M. (2001) Digital natives, digital immigrants. Part 2: Do they really think differently?, *On the Horizon*, **9** (6), 1-6.

Seth, G. (2012) *Analyzing the effects of social media on the hospitality industry*, master's thesis, University of Nevada, Las Vegas.

Sotiriadis, M.D. (2017) Sharing tourism experiences in social media, *International Journal of Contemporary Hospitality Management*, **29** (1), 179-225.

Ting, P.H., Wang, S.T., Bau, D.Y. & Chiang, M.L. (2013) Website evaluation of the top 100 hotels using advanced content analysis and eMICA model, *Cornell Hospitality Quarterly*, **54** (3), 284-293.

Tuten, T.L. (2020) *Social Media Marketing*. Sage.

Van Dijck, J. (2013) *The Culture of Connectivity: A critical history of social media*. Oxford University Press.

Verma, R., Stock, D. & McCarthy, L. (2012) Customer preferences for online, social media, and mobile innovations in the hospitality industry, *Cornell Hospitality Quarterly*, **53** (3), 183-186.

Yang, X., Pan, B., Evans, J.A. & Lv, B. (2015) Forecasting Chinese tourist volume with search engine data, *Tourism Management*, **46**, 386-397.

Zarrella, D. (2009) *The Social Media Marketing Book*. O'Reilly Media, Inc.

Zeng, B. & Gerritsen, R. (2014), What do we know about social media in tourism? A review, *Tourism Management Perspectives*, **10**, 27-36.

Ziyković, R., Gajić, J. & Brdar, I. (2014) The impact of social media on tourism, *Singidunum Journal of Applied Sciences*, 758-761.

2 Thematic social media research trends in hospitality and tourism:

A bibliometric approach 2019-2020

Marwa Al-Busaidi and Khaldoon Nusair

Introduction

Social media plays a significant role on travel decisions as travelers interact with others, provide reviews, and share experiences. As such, social media research is becoming increasingly popular in the hospitality and tourism contexts. Bibliometric analysis provides a retrospective review of published articles on a particular topic and suggests thematic trends based on this review. In recent years, more bibliometric studies have investigated the progression in social media publications and have suggested future research avenues on this topic (Nusair *et al.*, 2019; Chu *et al.*, 2020; Nusair, 2020). The increase in the total number of social media publications calls for more bibliometric research in order to find out the state-of-the-art research direction (Mehraliyev *et al.*, 2019). In this regard, this article utilized keyword and co-word analysis to examine the topical development in this area of research.

Despite the propagation of the social media topic, there are insufficient bibliometric studies that broadly integrate the growing body of articles published in this discipline (Nusair *et al.*, 2019). The increase in social media research signifies a robust rationale to perform bibliometric research to project patterns and paradigm changes in recent years. Past bibliometric studies have identified the major themes of social media research that have emerged over the past two decades. As a matter of fact, most bibliometric articles were limited to articles published prior to 2019. However, in the two years 2019-2020, research in this field has witnessed tremendous thematic progress in theory and practice. While it is challenging to identify progress on a yearly basis (Kapoor *et al.*, 2018), this study utilized bibliometric analysis to reveal social media topical advancement in those two years. As such, this chapter offers a broad analysis of the recent social media literature combining co-word analysis and thematic keyword analysis between 2019-2020. The chapter concludes with proposals to the most recommended future topics for social media research in hospitality and tourism.

Specifically, the objectives of this research are to investigate co-word analysis of social media research between 2019 and 2020, examine whether keywords effectively provide a summary of the content of the article, and propose topical trends to guide future social media research.

Literature review

An increasing number of bibliometric research studies have been published over the past twenty years in the domain of hospitality and tourism (Leung & Law, 2007; McKercher, 2008; Bowen & Whalen, 2017; Mariani *et al.*, 2018; Mirzaalian & Halpenny, 2019; Nusair *et al.*, 2019; Nusair 2020). These studies provided overviews of topical developments in the hospitality and tourism context. For example, Bowen and Whalen (2017) conducted a review to assess the current trends which shape the tourism industry. Their study revealed several topics including 'social media' as an emerging trend in the same context. However, following a systematic literature review approach, it was found that there is a lack of bibliometric analysis papers on social media in the hospitality and tourism domain; a relatively limited number of publications were identified (McKercher, 2008; Leung *et al.*, 2013; Zeng & Gerritsen, 2014; Leung *et al.*, 2017; Lu *et al.*, 2018).

For example, Leung *et al.* (2013) conducted content analysis and reviewed 44 social media publications in hospitality and tourism studies published between 2007 and 2011 to examine the impact of social media studies. Moreover, Zeng and Gerritsen (2014) identified a wide variety of research topics and offered a broader view of social media and its influence on tourism. By analyzing 279 articles from different sources published between 2007 to 2013, the results revealed 'user-generated content' to be the most widely used term in social media studies. Furthermore, additional investigations were conducted to discover a larger scale of evolving trends in social media published articles. The first research paper in social media to use co-citation and co-word analyses was Leung *et al.* (2017) study which covered a 9-year time span (2007-2016). A systematic bibliometric analysis was conducted to address the influence of social media using 406 articles. The results revealed a rapid growth of 'social media' topics in the period of 2012-2016. Similarly, Lu *et al.* (2018) used literature review analysis method to chart the progress of research between 2004 and 2014 in this field. After analyzing 105 social media papers published in seven journals, the report illustrated that social media research is progressing, yet remains in its initial stages, which calls for additional attention from social media scholars.

Notably, the evolving growth of social media research has recognized the need for more literature reviews to evaluate the thematic trends in social media research and to identify new emerging topics. As a result, several additional systematic reviews were conducted (Nusair *et al.*, 2019; Mehraliyev *et al.*, 2019; Mirzaalian & Halpenny, 2019; Chu *et al.*, 2020; Nusair, 2020). A comprehensive approach to explore the future course of social media research has been provided by Nusair *et al.* (2019). Systematic literature reviews in past research focused on peer-reviewed journals, field scholars, universities, and collaborations in inter-country studies. The study concluded with new emerging themes of research between 2011 to 2016 including: 'big data', 'netnography', 'Travel 2.0' and 'Web 2.0'. In addition, Mehraliyev *et al.* (2019) explored the growth in social media studies by conducting authorship and co-authorship analyses of 385 published papers. The study analyzed the most productive authors, their academic qualifications, regional differences, and universities networks. Furthermore, the Mirzaalian and Halpenny (2019) study is the first systematic literature review that applies social media analytics. This article reviewed hospitality and tourism studies using social media analytics collecting articles from five international electronic bibliographic databases from the previous 18 years. The findings revealed that a significant increase in social media literature reviews was noticeable since 2014. The study suggested an expansion of several approaches used in the same context.

Additionally, few recent systematic reviews of social media studies have been conducted in hospitality and tourism (Chu *et al.*, 2020; Nusair, 2020). For instance, Chu *et al.* (2020) is considered the first literature review paper on social media that covered the role of social media advertising in tourism research, provided an evaluation of the current literature, and suggested an avenue for future research in this discipline. It covered a 15-year timeframe (2004–2019) analyzing a collection of 192 articles based on eight main variables: year of publication, journal, country studied, research topic, social media types. Finally, Nusair (2020) provided a comprehensive bibliometric analysis to expose the fast growth in social media publications between 2002 and 2018. This study, in particular, reported the thematic trends in this area of research, by conducting co-word network analysis and thematic keyword analysis using a large sample of 601 studies. The study marked social media research between 2007-2012 as the beginning of a promising new area of research. Emerging trends were identified in the period (2013-2018) such as 'big data', 'e-tourism', 'green experience' and 'smart tourism'. Also, some interesting keywords related to social media emerged in the maturity stage along with social media platforms such as 'Expedia', 'Foursquare', 'Flickr', 'Pinterest', 'Couchsurfing' and 'Twitter'.

Drawing on previous literatures, there has been a noticeable growth in social media studies, however, there is a call for in-depth analysis of literature in the context of social media. In addition, several approaches were adopted in the past research, but none has examined how keywords provide a general summary of the articles' content. Thus, this study performed a bibliometric analysis and used co-word analysis and thematic keyword analysis for the purpose of reviewing social media literature progression in the tourism and hospitality domain. It is the first article to examine whether articles' keywords reflect the content of journal articles, by reviewing 276 published social media articles from hospitality and tourism journals over the two years (2019-2020). The study also aims to offer an outlook of social media trends in the years between 2019 and 2020 and an agenda for future research in the hospitality and tourism context.

Methodology

This study identified publications on the topic of social media in hospitality and tourism. To this end, a comprehensive systematic literature review was performed. In January 2021, the database of Australian Business Deans Council (ABDC, *version 2019*) was searched to identify social media publications between 2019 and 2020. The ABDC database includes a wider range of publications compared to other databases (Mura & Sharif, 2016). A total of 44 keywords from previous social media bibliometric publications were selected to spot articles (Nusair, 2020). Our search results reported 275 social media articles published in 39 journals. A research associate checked the validity of the search outcome.

Coding was completed in four steps:

1. The references of all articles were moved to RefWorks, then were transferred to a spreadsheet;
2. All categories (columns) were given distinct labels;
3. Soft copies of all publications were downloaded in order to get the details for each columns;
4. A research associate inputted data in the spreadsheet.

Results

Figure 2.1 illustrates the trend of social media research in the hospitality and tourism domain for both 2019 and 2020. A significant growth was observed between the two years, with nearly 176 published articles in 2020. This indicates that due to its important role, social media is gaining enough care from interested researchers in the tourism and hospitality domain.

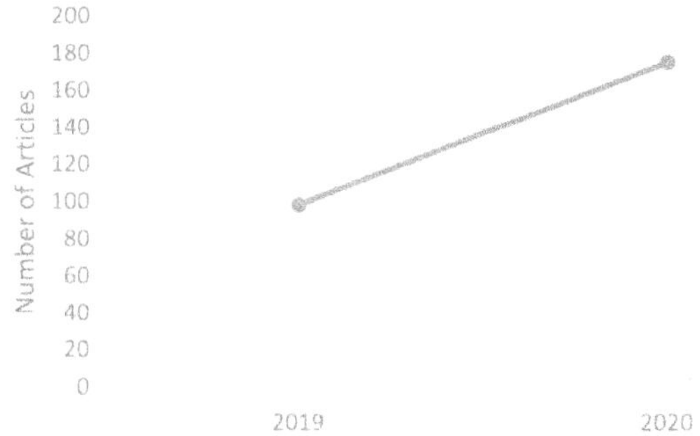

Figure 2.1: Progression of social media research in hospitality and tourism

Social media research trends in the leading journals

Social media research's progression was most marked in the five leading hospitality and tourism journals: *Current Issues in Tourism, International Journal of Contemporary Hospitality Management, Tourism Management, Journal of Travel Research and Journal of Travel* and *Tourism Marketing*. The aforementioned journals make up about 45 per cent of the total number of articles on social media between 2019-2020 (Appendix A). *Current Issues in Tourism* shows a remarkable growth specifically in 2020 and published around 37 articles (see Figure 2.2). *International Journal of Contemporary Hospitality Management* published around 27 articles. *Tourism Management published* around 24 articles. *Journal of Travel Research and Journal of Travel* published nearly 24 articles. Finally, *Tourism Marketing* published around 14 articles. It was observed that the three journals of *International Journal of Contemporary Hospitality Management, Tourism Management* and *Journal of Travel and Tourism Marketing* kept their position being part of the five leading journals that contribute to the social media research based on the findings of Nusair (2020).

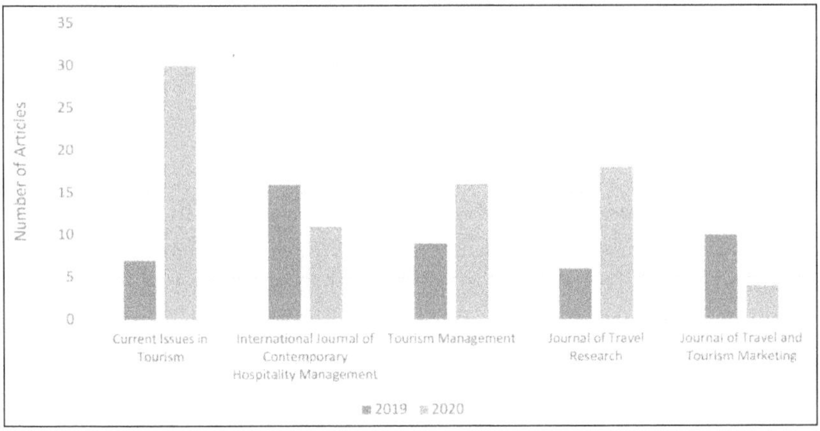

Figure 2.2: Evolution of social media research in top journals between 2019 and 2020

Co-word network analysis of social media research

To address the first objective, Figure 2.3 displays the clusters emerged in social media research between 2019 and 2020. The size of the bubbles indicates the content frequency, while the thickness of the lines indicates link strength. The analysis was created using a collection of keywords from the published articles in tourism and hospitality journals.

Five main clusters were developed, namely: 'social media', 'travel', 'marketing', 'trust' and 'customer satisfaction'. The first cluster is 'social media' which is linked to 'word of mouth', 'eWOM' and 'big data' in 'hospitality' and 'tourism' industries. It was observed that 'Twitter', 'Facebook' and 'TripAdvisor' are the main studied platforms in the last two years. The second cluster 'marketing' linked with 'destination marketing', 'destination image' and 'reputation'. The third cluster is 'travel' industry which is mainly connected with 'satisfaction', 'expectations', 'consumption' and 'motivations'. Another cluster is 'trust' which is recognized to be related to 'privacy', 'credibility' and 'destination brand' literatures. Finally, 'customer satisfaction' which is mainly linked to 'service failure' and 'service recovery'.

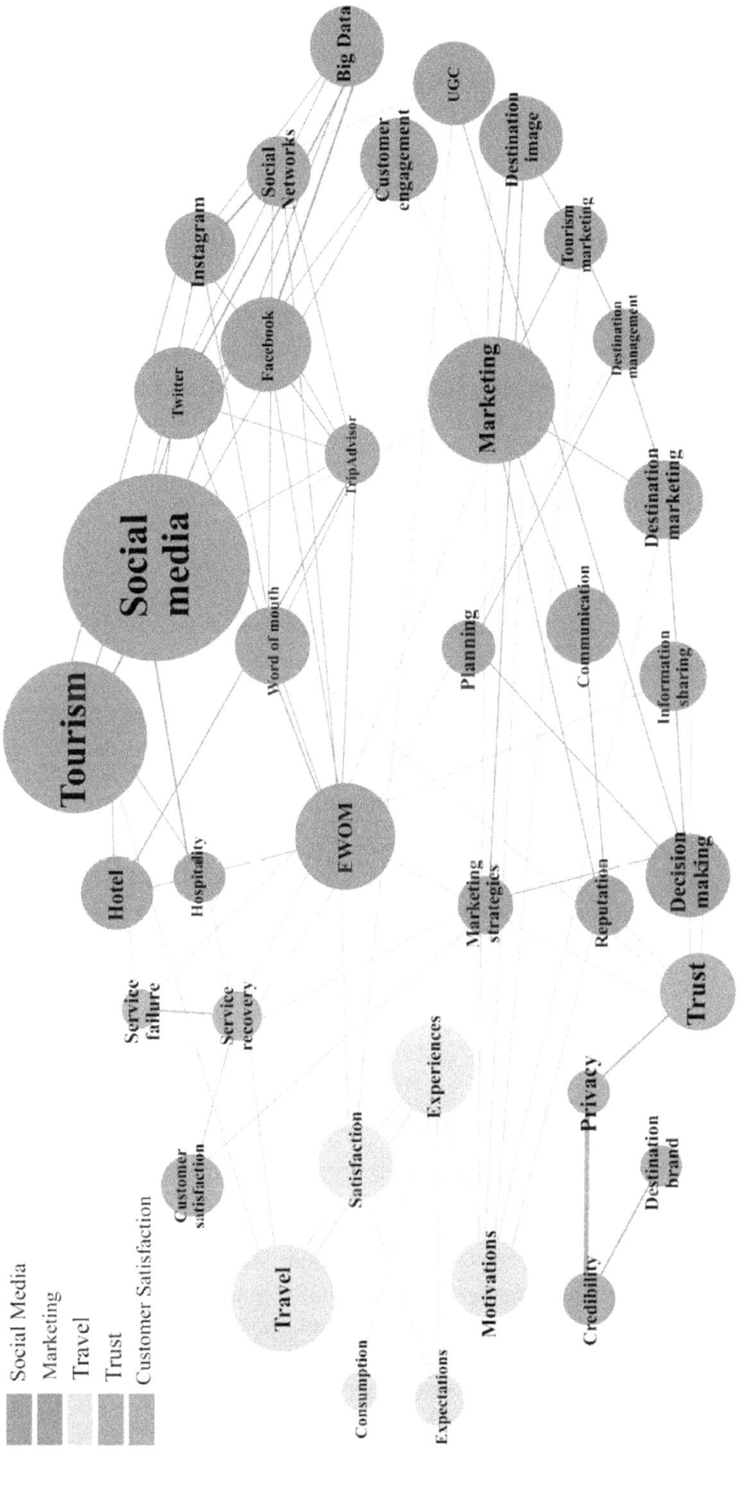

Figure 2.3: Co-word network of social media research in tourism and hospitality 2019-2020

Keyword comparison: Listed in the abstract and implications

To address the second objective, a test was run for 2020 articles to identify the degree of similarity and reflection between keywords listed in the abstract and the implication keywords. Figure 2.4 illustrates the similarities between the articles' keywords and the extracted keywords from the implications; the area with the spike represents keywords listed in the abstract and the rounder area represents extracted keywords from implication. The intersecting area between the two defines the degree of similarity between both keywords. Table 2.1 showcases the common and uncommon thematic keywords between the two. The results provided a strong evidence that thematic keywords effectively provide a summary of the content of the article as 72 per cent of the keywords were in common.

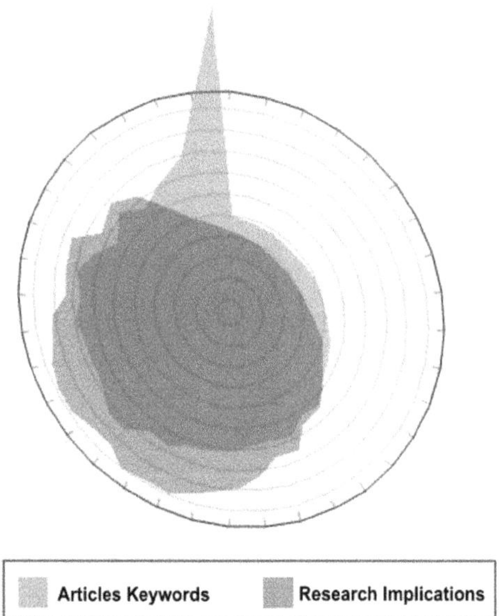

Figure 2.4: Keyword comparison: listed in abstract and implications

Table 2.1: Common and uncommon keywords between articles' keywords and their content. Similarity = 72 (%)

Common keywords
social media, tourism, marketing, travel, ewom, experiences, destination image, customer engagement, decision making, destination marketing, big data, hotel, satisfaction, Instagram, customer satisfaction, destination management, tourism marketing, consumer reviews, online communities, marketing strategies, customer involvement, destination brand, electronic word of mouth, user generated content, digital marketing, social network, networking sites, tourism management, user generated, word of mouth, brand engagement (**N=31**)

Uncommon keywords

Twitter, information sharing, consumer reviews, TripAdvisor, knowledge management, internet marketing, content analysis, online reviews, service recovery, consumer behavior, hospitality, sentiment analysis. (**N=12**)

Recommended future research topics in social media studies

To meet the third objective, the packed bubble chart in Figure 2.5 visualizes the stated topics that require further attention and social media research in the hospitality and tourism industries, based on 2019-2020 articles. The size and shade of the bubbles represent total frequency. The top seven topics recommended for further explorations and research are: 'motivations', 'perceptions', 'social media', 'marketing', 'big data', 'destination image' and 'tourist experience'.

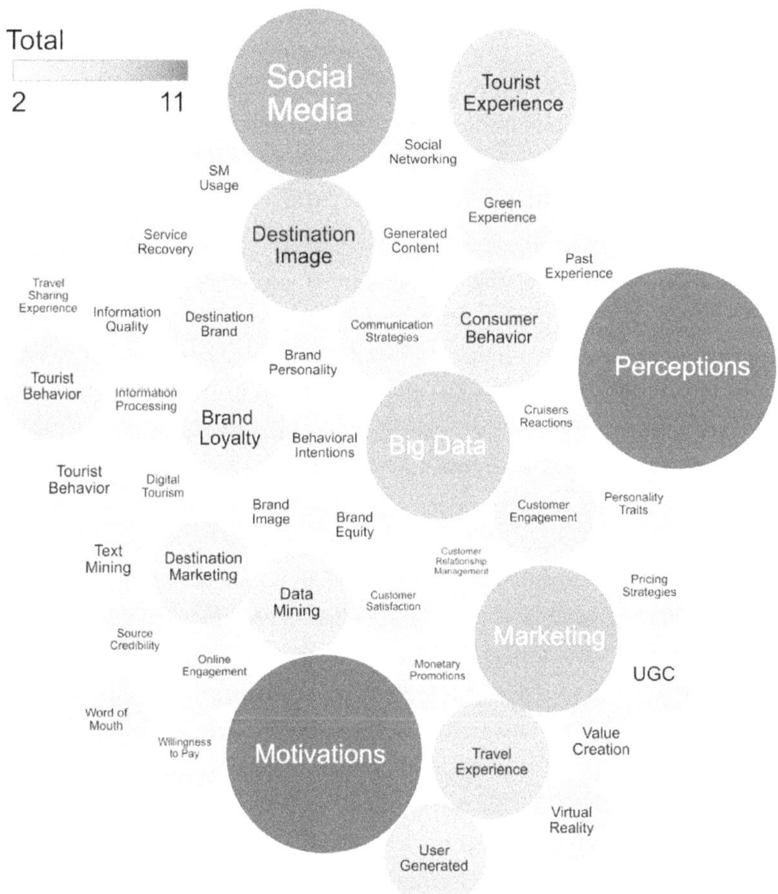

Figure 2.5: Most recommended future topics for social media research in hospitality and tourism

Conclusion

This chapter attempted to provide a literature review analysis to assess social media research progression in hospitality and tourism. It conducted a bibliometric analysis of 275 social media articles published in 39 hospitality and tourism journals between 2019 and 2020. The main contribution of this article centered on three main objectives: to perform co-word network analysis of social media studies published between 2019 and 2020; to explore whether an article's keywords effectively provide an overview of the content of the article; to propose thematic trends on social media to guide future research.

It was observed that there had been a remarkable growth in a short period of time (2019-2020) in terms of the number of published articles. In addition, five top journals were identified to be contributing greatly to social media research: *Current Issues in Tourism* is the at the top and it indicates that social media is currently gaining additional spotlight from scholars and researchers in hospitality and tourism fields. Keywords were extracted and literature evaluation measures were implemented to identify social media topical trends, such as: co-word network analysis and thematic keyword analysis.

One main objective of this article was to implement co-word network analysis of social media research published between 2019 and 2020 to define the conceptual structure of social media research. Five main social media co-word clusters were recognized in the period of 2019-2020 which indicates that social media is continuing to gain additional studies in various fields and different approaches. Similar to the findings of Nusair's (2020) study, social media is starting to gain additional diversity with an increase in number of trended topics throughout the years; and contradicts with Lu *et al.*'s (2018) study that stated social media studies lack variation. The analysis revealed several relationships with 'social media', such as: 'word of mouth' 'eWOM', and 'big data' in mainly 'tourism' and 'hospitality' industries. Notably, Facebook, Twitter, TripAdvisor and Instagram are the most studied platforms based on the collected articles. Facebook is the most frequently studied social media platform which agrees with the findings of Chu *et al.* (2020). This can be used for future studies that might, for example, investigate why Facebook is gaining extra attention from scholars and how it is serving the tourism and hospitality industries, and how other platforms can benefit from these studies. Moreover, the findings of Zeng and Gerritsen (2014) determined 'marketing' as the most dominant research focus, providing an overview for future studies in social media research.

Our study clarified that 'marketing' is abundantly present in the current articles of social media in hospitality and tourism context, which indicates that scholars and researchers have directed their attention and shed light on the topical area. For example, based on our findings, there is a link between 'marketing' and other topics, such as: 'destination marketing', 'destination image' and 'reputation'. Our finding also found that 'satisfaction' is gaining wider attention from scholars in social media research; 'travel' studies have examined travel 'experiences', tourists 'consumptions' and their 'motivations' in relation to 'satisfaction'. In addition, 'customer satisfaction' is used as an essential measure to assess 'service failure' which contributes in providing strategies for 'service recovery'. Additionally, it is interesting to note that 'trust' is gaining much attention from scholars in social media studies. Nusair (2020) pointed out that 'trust' is in the growth stage as it is one of the topics which receives sufficient attention from scholars. Our findings illustrate that 'trust' is linked to 'privacy', 'credibility' and 'destination brand' literatures.

Another contribution of this study was to explore whether listed keywords effectively provide an overview of articles' content. Our study collected keywords from articles and also examined content by gathering keywords from implications. The process was done by taking the most frequent keywords and comparing their presence in each of articles' keywords and implications. Different common keywords were identified such as 'social media', 'tourism', 'marketing', 'travel', 'eWOM', 'experiences', 'destination image'. The results provided a strong evidence that thematic keywords effectively provide a summary of the content of the article as 72 per cent of the keywords were in common. To validate our findings, more future research can be conducted over a larger time span.

Finally, to meet the third objective, several future topical themes were identified in this study. The suggested topical themes were visualized based on total frequency. For 2019-2020 social media articles, the top seven topics that require additional attention from scholars and could provide opportunities for future research are: 'motivations', 'perceptions', 'social media', 'marketing', 'big data', 'destination image' and 'tourist experience'. Future studies might, for example, take one of these topics and perform more in-depth analysis to define areas of opportunities and explore any unstudied links that could benefit the domain of social media research.

Discussion questions

1. Provide two indicators that social media research has been gaining considerable attention from scholars between 2019-2020.
2. This research highlighted 'satisfaction' as a widely studied concept in social media research, in your opinion how has 'satisfaction' changed the hospitality and tourism industries?
3. Discuss the methodological procedures needed to conduct co-word and thematic keyword analysis in social media research in hospitality and tourism.
4. How can co-word analysis create opportunities for future research on social media research in hospitality and tourism? Explain your answer.

References

Bowen, J. & Whalen, E. (2017) Trends that are changing travel and tourism, *Worldwide Hospitality and Tourism Themes*, **9** (6), 592-602.

Chu, S. C., Deng, T. & Cheng, H. (2020) The role of social media advertising in hospitality, tourism and travel: A literature review and research agenda, *International Journal of Contemporary Hospitality Management*, **32** (11), 3419-3438.

Kapoor, K.K., Tamilmani, K. & Rana, N.P. (2018) Advances in social media research: Past, present and future, *Information Systems Frontiers*, **20** (3), 531–558.

Leung, R. & Law, R. (2007) Analyzing research collaborations of information technology publications in leading hospitality and tourism journals: 1986–2005, *Information and Communication Technologies in Tourism*, 547-556.

Leung, D., Law, R., Van Hoof, H. & Buhalis, D. (2013) Social media in tourism and hospitality: A literature review, *Journal of Travel & Tourism Marketing*, **30** (1-2), 3-22.

Leung, X.Y., Sun, J. & Bai, B. (2017) Bibliometrics of social media research: A co-citation and co-word analysis, *International Journal of Hospitality Management*, **66**, 35-45.

Lu, Y., Chen, Z. & Law, R. (2018) Mapping the progress of social media research in hospitality and tourism management from 2004 to 2014, *Journal of Travel & Tourism Marketing*, **35** (2), 102-118.

Mariani, M., Baggio, R., Fuchs, M. & Höepken, W. (2018) Business intelligence and big data in hospitality and tourism: A systematic literature review, *International Journal of Contemporary Hospitality Management,* **30** (12), 3514-3554.

McKercher, B. (2008) A citation analysis of tourism scholars, *Tourism Management,* **29** (6), 1226-1232.

Mehraliyev, F., Choi, Y. & Koseoglu, M.A. (2019) Social structure of social media research in tourism and hospitality, *Tourism Recreation Research,* **44** (4), 451-465.

Mirzaalian, F. & Halpenny, E. (2019) Social media analytics in hospitality and tourism: A systematic literature review and future trends, *Journal of Hospitality and Tourism Technology,* **10** (4), 764-790.

Mura, P. & Sharif, S.P. (2017) Narrative analysis in tourism: A critical review, *Scandinavian Journal of Hospitality and Tourism,* **17** (2), 194-207.

Nusair, K. (2020) Developing a comprehensive life cycle framework for social media research in hospitality and tourism, *International Journal of Contemporary Hospitality Management,* **32** (3), 1041-1066.

Nusair, K., Butt, I. & Nikhashemi, S.R. (2019) A bibliometric analysis of social media in hospitality and tourism research, *International Journal of Contemporary Hospitality Management,* **37** (7), 2691-2719.

Zeng, B. & Gerritsen, R. (2014) What do we know about social media in tourism? A review, *Tourism Management Perspectives,* **10** (2014), 27-36.

Appendix A: Journals and articles' count

Journal	Articles
Current Issues in Tourism	38
International Journal of Contemporary Hospitality Management	27
Tourism Management	25
Journal of Travel Research	24
Journal of Travel and Tourism Marketing	14
Journal of Hospitality and Tourism Technology	11
Tourism Management Perspectives	11
Journal of Hospitality and Tourism Management	10
Journal of Sustainable Tourism	10
Asia Pacific Journal of Tourism Research	9
Journal of Hospitality Marketing and Management	9

e-Review of Tourism Research	8
Information Technology and Tourism	8
International Journal of Hospitality Management	8
International Journal of Hospitality and Tourism Administration	7
Journal of Hospitality and Tourism Research	6
Tourism Recreation Research	6
International Journal of Culture, Tourism, and Hospitality Research	5
International Journal of Tourism Research	5
Journal of China Tourism Research	5
Journal of Hospitality and Tourism Insights	5
Worldwide Hospitality and Tourism Themes	5
Cornell Hospitality Quarterly	4
Tourism Geographies	4
International Journal of Tourism Cities	3
Journal of Tourism and Cultural Change	3
Journal of Tourism Futures	3
Tourism and Hospitality Research	3
Tourism Economics	3
Annals of Tourism Research	2
Journal of Quality Assurance in Hospitality and Tourism	2
Tourism Review	2
International Hospitality Review	1
International Journal of Tourism Policy	1
Journal of Hospitality and Tourism Education	1
Journal of Hospitality, Leisure, Sport and Tourism Education	1
Journal of Human Resources in Hospitality and Tourism	1
Journal of Policy Research in Tourism, Leisure and Events	1
Tourism Planning and Development	1

3 Social media marketing activities: Studying social identification, trust, value and loyalty

Faizan Ali, Abraham Terrah, Seden Dogan and Cihan Cobanoglu

Introduction

Social media refers to different online platforms and tools that provide social interaction between users (Hays *et al.*, 2013) through text, pictures, and videos (Berthon *et al.*, 2012). During the last decade, the number of social media users has increased rapidly. The latest report published by We Are Social (2021) shows that in 2020 there were currently 4.2 billion social media users, equivalent to more than 53% of the world's total population. In 2019, this figure was 3.8 billion (We Are Social, 2020), showing an increase of more than 13% over the year. In other words, social media gained 1.3 million new users every day, or 15 people every second during 2020. A typical user spends almost 2.5 hours in a day on social media (We Are Social, 2021). They use social media for different purposes, including fear of missing out, need of fulfillment (Aksoy, 2018), search for information, search for social interaction (Chen & Lin, 2019), search for positive emotions, and escape from negative emotions (Brailovskaia *et al.*, 2020).

Social media is also becoming a popular marketing platform for brands (Chaffey, 2020; Wang *et al.*, 2012). Social media brings significant opportunities for hospitality brands to engage and communicate with their consumers (Chu *et al.*, 2020). It has created a transformation in consumers' behavior where consumers can interact with brands during different stages of purchase decisions, such as discovering a brand or product, looking for information about a brand or product, and looking for suggestions from peers about a brand or product. Moreover, it allows consumers to exchange information with one another, thus contribute to enhancing electronic word-of-mouth by (Kaosiri *et al.*, 2019), giving birth to social groups where the members have a good sense of community (Zhou *et al.*, 2020). These online communities facilitate customer-customer and customer-brand interactions, which can shape potential customers' perceptions and ideas. Also, hospitality brands have started to use social media influencers, who are people who have built a reputation for their knowledge and expertise on a specific topic and gained a big number of followers (Bakker, 2018). Through endorsement by social media influencers, brands promote their products/services on social media platforms including Instagram, TikTok, Facebook, Youtube, and Weibo (Xu & Pratt, 2018). It provides brands the opportunity to establish individual connections, with a chance to reach out to the customers in a much more personalized way (Seo & Park, 2018).

Owing to these benefits, social media has incontestably become an essential component for hospitality brands' online marketing strategies (Pham & Nguyen, 2019). Social media marketing refers to marketing practices on social media platforms conducted by companies to influence users' behavior (Chen & Lin, 2019). According to Kim and Ko (2012), social media marketing activities (SMMA) include entertainment, interaction, trendiness, customization, and word-of-mouth (WOM). Similarly, Yadav and Rahman (2018) also confirmed interactivity, informativeness, word-of-mouth, trendiness, and personalization as dimensions of SMMA. In this study, the SMMA components proposed by Kim and Ko (2012) are studied in the context of social media advertising by hospitality brands. It is important to understand the role of SMMA in creating value perceived by the customer, which is an essential output of marketing activities (Sanchez *et al.*, 2006). Moreover, hospitality brands need to build strong relationships with their customers to maintain trust and loyalty. Trust is a crucial element to building a sustainable relationship with customers (Wang *et al.*, 2014). Loyalty, on the other hand, helps to develop long-term relationships, which leads to repeated purchase behavior and intention to pay more (Kandampully *et al.*, 2015). Although previous studies regarding social media focused on SMMA's impact on social identification and perceived value (Chen & Lin, 2019), studies focused on developing loyalty through social identification, perceived value, and trust are very limited. The purpose of this study is to fill this deficiency and address the following question: how SMMA performed by the hotels may develop social identification, trust, value, and loyalty towards the hotel.

Literature review

The rise of social media networks in the 2010s has attracted marketers' attention, who view the platforms as powerful advertising tools due to the reach in terms of a number of consumers. For instance, Facebook counted over 2.7 billion active users and YouTube 2 billion users as of October 2020 (Statista, 2020). TikTok, another rising social network especially in the United States, has about 689 million users (Statista, 2020). The spectrum of means to market online using social media is broad, including opportunities to share information through blogs, forums, or virtual communities. As such, several organizations seek to extend the levels of communication and interactions with consumers in order to create brand communities on the virtual level (sphere) (Kaplan & Haenlein, 2010; Chen *et al.*, 2014). According to Wellman (2001), consumers' needs in terms of online communities do not differ from the ones in offline communities, especially in terms of shared experiences, social support, and self-identity. Previous studies have established that in a brand community, a set of beliefs develops among the members, and eventually links them by sharing their experiences with the brand (Muniz & O'Guinn, 2001; Muniz & Schau, 2005; Sicilia & Palazon, 2008). This implies that consumers may also be interested in creating experiences through interactions with other consumers or experts (Prahalad & Ramaswamy, 2004).

Social media marketing activities

The social media marketing tools operate through online communities created around certain brands and organizations, bringing together individuals having experiences with those brands and organizations. An online brand community can be defined as a community in which interactions occur through computer-mediated devices and in which members share similar interests (Baldus *et al.*, 2015). Online brand communities have been enhanced by social media, which provides a way for two-way communication and allows the sharing of content between brands and consumers on the one hand, and between consumers themselves on the other hand (Verhagen *et al.*, 2015). Yadav and Rahman (2017) say that social media marketing tools allow for interaction, sharing of crucial information, and the spread of word-of-mouth about specific products and services. These are made possible through elements such as user ratings, reviews, referrals, testimonies, as well as recommendations (Yadav & Rahman, 2017). As a consequence, social media gained in importance, in particular with regard to managing relationships with consumers (Hollebeek *et al.*, 2014). A definition of social media marketing is provided by Yadav and Rahman:

> *A process by which companies create, communicate, and deliver online marketing offerings via social media platforms to build and maintain stakeholder relationships that enhance stakeholders' value by facilitating interaction, information sharing, offering personalized purchase recommendations, and word of mouth creation among stakeholders about existing and trending products and services (2017: 1296).*

Yadav and Rahman (2017) provided in their study a five-dimension scale of social media marketing activities, namely interactivity, informativeness, word-of-mouth, trendiness, and personalization. Prior to those findings, Kim and Ko (2012) also provided—in their investigation of luxury brands—five dimensions to social media marketing activities. These include entertainment, interaction, trendiness, customization, and word-of-mouth. Ebrahim (2020) found three dimensions to social media marketing activities: trendiness, customization, and word-of-mouth, and established their link with loyalty. Working from the prior literature, the present study argues social media marketing entails performing activities that capture and engage consumers via five distinct factors—entertainment, interaction, trendiness, customization, and word-of-mouth (Chen & Lin, 2019; Godey *et al.*, 2016; Kim & Ko, 2012; Mohammadpour *et al.*, 2014; Yadav & Rahman, 2018). The entertainment dimension is hedonic and denotes the pleasure derived from the use of social media. Interaction allows users to contribute to the brand; trendiness is concerned with the novelty and up-to-dateness of the information; customization relates to the ability to serve different types of needs and consumer demands; and word-of-mouth denotes opinions and experiences' information sharing between members.

Social Identification

Social identity theory (Tajfel & Turner, 1986) has previously been used as a theoretical foundation for understanding brand identification in marketing (Elbedweihy *et al.*, 2016; Lam *et al.*, 2013; Rather & Sharma, 2017). According to this theory, consumers spend a considerable amount of effort developing their own social identity (Bhattacharya & Sen, 2003; Islam *et al.*, 2017). They tend to develop a group affiliation, which allows them to craft a unified opinion that acts as a determinant of a favorable attitude (Shareef *et al.*, 2019). When consumers can communicate with their peers and share their experiences with them, they are also provided with the capability to develop collective decisions. Thus, social identification is the element that allows consumers to be engaged with a particular brand or organization (Hollebeek *et al.*, 2016; Sprott *et al.*, 2009; Turner *et al.*, 1987), through their developed social identity and group affiliation. In this equation, marketers mainly stress the similarities between members and dissimilarities with non-members (Hollebeek, 2018). Thus, members find themselves positively identifying with the brand. Lam *et al.* (2013: 235) defined brand identification as a *"psychological state of perceiving, feeling, and valuing his or her belongingness to a brand"*. As consumers identify with brands, they develop attitudes related to those specific brands (Stokburger-Sauer *et al.*, 2012), mostly because of brands' social media marketing activities (Chen & Lin, 2019). Given the importance for brands of social media for brands to facilitate social identity expression and development, we posit the following hypothesis:

H1: Social media marketing activities have a positive effect on social identification.

Perceived value

The concept of value represents a very important factor in determining consumers' choices and behaviors. A consumer's loyalty toward a brand or organization is increased as the perceived value derived from that consumer's overall evaluation is greater (Sirdeshmukh *et al.*, 2002). Similarly, Farquhar and Rowley (2006) advanced that consumers' loyalty with respect to a social community is a function of the value created by this community and as perceived by the consumer. Perceived value is best understood as a psychological evaluation that is not only derived from the product features but also cognitive evaluations from the consumers and regarding the products (Tynan *et al.*, 2010). In that sense, consumer value can better understand how members might evaluate a particular social community. With regards to social media, several researchers have turned their attention to advertisement and viral marketing (e.g., Barnes & Mattson, 2009; Bonds-Raacke & Raacke, 2010; Schudson, 2013), arguing that discourse and hedonic enjoyment—which is concerned with entertainment—were very important in shaping value perceptions of customers. Shareef *et al.* (2019) has found in their study in the context of advertisement that entertainment was one the major forces leading a consumer to perceive a higher value of an advertising product, especially when peer reviews are positive. Previous studies have established that advertising is one of the initial activities from a brand that can determine customer's perceived value (Di Pietro & Pantano, 2012; Hughes & Palen, 2009; Logan *et al.*, 2012; Shareef *et al.*, 2015). In the context of marketing activities performed on social media, we posit:

H2: Social media marketing activities are positively related to perceived value.

Trust

In the reign of new information technologies, trust remains a fundamental issue and investigation domain by researchers. Marketers are often challenged with enticing trust within their customers in online settings; this can be attributed to the non-existence of face-to-face communication, but also to the high propensity of user-based content generation (PWC, 2018; Yahia et al., 2018). For instance, in online shopping, Pakistani consumers' most common complaints were unmet expectations, faulty products, delivery delays, overstated prices, etc. (GallupPakistan, 2019). As compared to the real world, trust in the online settings can be viewed as a substitute for experiential cues (e.g., physical product, salesforce) (Verma *et al.*, 2016). According to Chatterjee (2001), consumers display more trust and perceive more value for information they received from their peers than from marketers. This implies that consumers would be more reliant on responsive peers, willing to share information and personal experiences. It is now common practice for young consumers in the Asia Pacific Region to look out for reviews or information pieces created by their peers when evaluating online content's credibility (eMarketer, 2016). This follows that it becomes vital for marketers to have their promotional messages shared between peers of social network and celebrities, brand ambassadors, and experts. Transactions or activities performed over social media also require trust in the operators to be successful, suggesting a positive correlation between trust and social media marketing activities. Therefore, we posit:

H3: Social media marketing activities are positively related to trust.

Loyalty

Brand loyalty can be defined as a deeply held commitment to rebuy a product or service in the future, regardless of situational factors (Chaudhuri & Holbrook, 2001). Several researchers focused on the antecedents and drivers of loyalty (e.g., Khadim *et al.*, 2018; Srinivasan *et al.*, 2002; Tatar & Eren-Erdogmus, 2016). Nevertheless, there is a lack of consensus on optimizing loyalty factors, which can be explained by the variations according to the context. For instance, in social media marketing activities, consumers' loyalty can be enhanced with updated a nd trendy information (Ebrahim, 2020). While certain studies focused on trust or perceived quality (Liat *et al.*, 2017; So *et al.*, 2013) as antecedents of loyalty, other examinations focused on perceived value. As related to perceived value, consumers perceiving high value from social media marketing activities are more likely to display loyalty to the brand or organization (Sirdeshmukh *et al.*, 2002). Regarding e-commerce, social media marketing activities with a positive perspective are likely to be enhance loyalty (Ismail, 2017). It is also important for brands and organizations to offer high-quality products and services that are in line with consumers' needs and values to improve their brand or social identification. As consumers identify with brands, they develop attitudes related to those specific brands (Stokburger-Sauer *et al.*, 2012). He *et al.*, (2012) explained that a consumer's social identification—through the social identity theory—correlates with loyalty as a consumer-based outcome. Also, Rather and Hollebeek (2019) found in their study that along with commitment and satisfaction, brand identification and trust had a positive impact on loyalty. Subsequently, with prior studies, the present study suggests that social identification, perceived value, and trust positively influence loyalty. Consequently, we propose the following hypotheses:

H4: Social identification has a positive influence on loyalty.

H5: Perceived value has a positive influence on loyalty.

H6: Trust has a positive influence on loyalty.

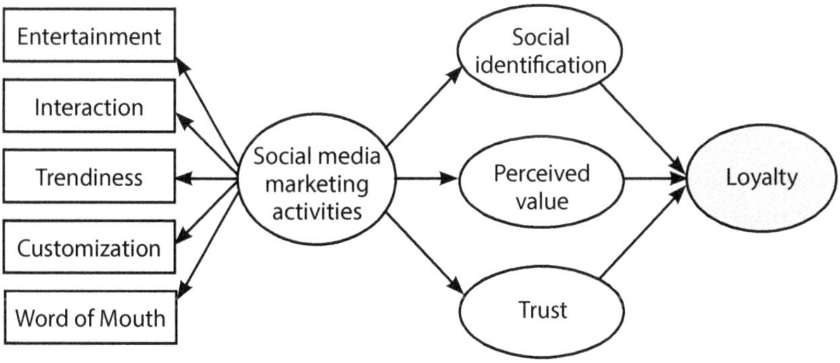

Figure 3.1: depicts the theoretical framework for this study.

Methodology

The study targeted users over the age of 18 who interact with hospitality-related brands via social media platforms. Amazon Mechanical Turk—which is an online survey platform and frequently used by hospitality and tourism researchers (Ali *et al.*, 2020)—was used for data collection. Due to the voluntary character of respondents' participation to this research, self-selection sampling was used as sampling method, appropriately with the purpose of the study. For data quality purposes, two attention checks questions were used; as well, the survey only included respondents who scored at least 95% Human Intelligence. As such, respondents that failed at correctly answering the attention checks, or provided incomplete responses, were removed from the final dataset. Consequently, further analysis was conducted on a total of 402 valid responses. Using Power analysis, this sample size satisfies the required sample size condition. Males constituted more than half of the respondents (53.7%). 40.2% of respondents were between the ages of 18 and 34. Around half of respondents (48.1%) were full-time employees, and approximately 60% of employees used more than three social media channels for their favorite hospitality brands.

Scales were adapted from prior research. We evaluated social media marketing activities using 11 items adapted from Kim and Ko (2012). Four items are used to assess brand trust (Tran & Strutton, 2020). Three items from Kwon and Wen (2010) are used to assess social identification. The perceived value is quantified using the three-item Liu *et al.* scale (2012). Loyalty is quantified using three criteria derived from Tran and Strutton (2020). A five-point Likert scale—ranging from strongly disagree to strongly agree—was used for all scales in this study, with the exception of loyalty, which was measured through a seven-point Likert scale. In order to account for common method variance, various statistical and methodological tools were employed. In fact, distinct cover stories were used for each measurement scale, in the optic of creating psychological separation between respondents. In addition, the 24 questions in the questionnaire were deemed brief enough, thus preventing confusion and fatigue so as to limit the negative effect on respondents' cognitive effort when answering the questions, and enhancing accuracy of answers. Third, independent and dependent variables were quantified using distinct measurement scales. Harman's single-factor test was performed, and the results indicated that no single factor explained the majority of the variance. As a result, common method bias was unlikely to have had a significant effect on this study.

Results

To validate the research model, we used Partial Least Squares based (PLS-SEM) analysis with smartPLS 3.0. Mardia coefficients did not indicate a multivariate normal distribution for the data, therefore, PLS-SEM was deemed most appropriate method for the research model to be tested (Ali *et al.*, 2018).

The coefficients of Mardia indicated that the data did not follow a multivariate normal distribution, which is why PLS-SEM was selected as the preferred method for testing the research model (Ali *et al.*, 2018). With smartPLS, two steps were taken to implement the analyses: measurement model analysis and structural model assessment. As suggested by Ali *et al.* (2018), reflective constructs were validated through composite reliability, Cronbach's alpha, and Average Variance Extracted (AVE). All the values passed the recommended thresholds, and, hence, each construct's reliability and validity was established (See Table 3.1). Moreover, discriminant validity was established through Fornell and Larcker (1981) and HTMT procedure.

Next, the structural model was assessed to assess the hypothesized relationships, with a bootstrapping size of 5000. The results are presented in Table 3.2.

Table 3.1: Validity and reliability

Items		Loadings	Cronbach's Alpha	rho_A	Composite Reliability	AVE
Entertainment						
E1	The content found in Brand X's social media seems interesting	0.914	0.713	0.732	0.856	0.749
E2	It is fun to collect information on products through Brand X's social media.	0.814				
Customization						
C1	It is possible to search for customized information on Brand X's social media.	0.834	0.704	0.714	0.825	0.701
C2	Brand X's social media provide customized services.	0.841				
Interaction						
I1	It is easy to convey my opinion through Brand X's social media.	0.876	0.823	0.825	0.895	0.739
I2	It is easy to convey my opinions or conversation with other users through Brand X's social media.	0.854				
I3	It is possible to have two-way interaction through Brand X's social media.	0.848				
Word of Mouth						
W1	I would like to pass information on products/services from Brand X's social media to my friends.	0.946	0.871	0.876	0.939	0.885
W2	I would like to upload content from Brand X's social media on my own social media.	0.935				
Trendiness						
T1	Using Brand X's social media is very trendy.	0.932	0.851	0.851	0.931	0.870
T2	The content on Brand X's social media is the newest information.	0.934				
Perceived Value						
V1	SMS advertising by Brand X is useful	0.748	0.831	0.858	0.899	0.750
V2	SMS advertising by Brand X is valuable	0.920				
V3	SMS advertising by Brand X is important	0.919				

Perceived Trust						
Tr1	Brand X social media advertising seems genuinely committed to my satisfaction	0.834	0.899	0.900	0.930	0.768
Tr2	Overall, I trust Brand X's social media advertising	0.919				
Tr3	In terms of usability, I know what to expect from Brand X's social media advertising	0.890				
Tr4	if Brand X makes a claim about in their social media advertising, it is probably true	0.860				
Social Identification						
SI1	As a member of Brand X's social media, my position is very important to me	0.890	0.908	0.911	0.942	0.845
SI2	As a member of Brand X's social media, I am the type of person who likes to engage in my community	0.926				
SI3	As a member of Brand X's social media, I feel their activities play an important part in my life	0.940				
Loyalty						
L1	I prefer Brand X's products/services over other alternatives	0.921	0.787	0.701	0.770	0.538
L2	I am loyal to Brand X	0.556				
L3	Brand X is my first choice for the product/service category they offer	0.676				

Table 3.2: Hypotheses Testing

Hypotheses		Beta	T-Value	P Values	Decision
H1	Social Media Marketing Activities -> Social Identification	0.783	30.978	0.000	Supported
H2	Social Media Marketing Activities -> Perceived Value	0.760	23.977	0.000	Supported
H3	Social Media Marketing Activities -> Trust	0.802	33.884	0.000	Supported
H4	Social Identification -> Loyalty	0.152	4.578	0.000	Supported
H5	Perceived Value -> Loyalty	0.133	3.408	0.000	Supported
H6	Trust -> Loyalty	0.461	4.318	0.000	Supported

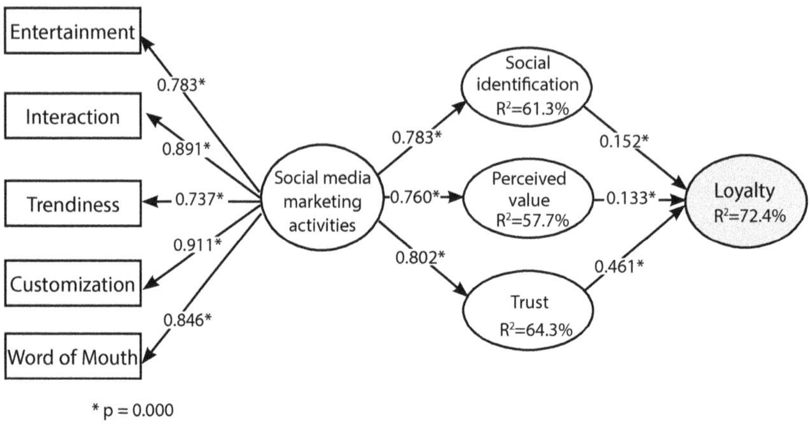

Figure 3.2: Structural Model

As seen in Figure 3.2, social media marketing activities on all five dimensions, namely entertainment, interaction, trendiness, customization, and word of mouth, are significant. Social media marketing activities were found to have a strong and positive influence on social identification (β=0.783, p<0.001), perceived value (β=0.760, p<0.001), and trust (β=0.802, p<0.001). Thus, the hypotheses H1 to H3 were supported. Moreover, loyalty was found to be strongly and positively influenced by social identification (β=0.152, p<0.001), perceived value (β=0.133, p<0.001), and trust (β=0.461, p<0.001). Thus, the hypotheses H4 to H6 were supported. Lastly, as shown in Figure 3.2, social media marketing activities explain 61.3%, 57.7%, and 64.3% variance in social identification, perceived value, and trust, respectively, whereas all three of these further explain 72.4% variance in loyalty.

Conclusion

The aim of this study was to examine the effects of SMMA on customers' social identification, trust, and value leading to customer loyalty towards hospitality brands. Data collected from regular social media users and followers of hospitality brands was subjected to PLS-SEM. The results show that SMMA have five significant characteristics: entertainment, interaction, trendiness, customization, and word of mouth. These results are in line with previous studies (Ebrahim, 2020; Kim & Ko, 2012; Yadav & Rahman, 2017). Moreover, findings indicated a significant and positive impact of social media marketing activities on social identification, perceived value, and trust. These findings also corroborate the findings from investigations by Chen and Lin (2019), Ebrahim (2020), and Zollo *et al.* (2020). Lastly, findings indicated that social identification, perceived value, and trust have significant and positive impacts on customer's loyalty with hospitality brands. These findings also support previous literature (Chen & Lin, 2019; Ebrahim, 2020; Rather & Hollebeek, 2018).

The present study draws theoretical and practical conclusions from these findings. One of the implications is that we added to the body of knowledge about social media marketing activities for hospitality brands. While numerous studies focused on SMMA in generic brands, financial service providers, and fashion brands, not enough investigations were conducted in the context of hospitality brands. Nevertheless, since hospitality brands are getting global recognition, identification, and growth (Zhou *et al.*, 2020), it is necessary to expand the scope of research into SMMA in the context of hospitality brands. Second, while many previous studies have focused on how SMMA can translate into purchase intentions, this study emphasized the role of customer's social identification, trust and value in between SMMA and loyalty. In fact, social media marketing as a mean of communication aims at involving consumers and providing them with useful information about brands. Therefore, it is important for users to identify themselves within the social circle, get value from all the activities by brands on social media, and provide continued trust in the face of data breaches and data privacy etc., (Ebrahim, 2020; Chen & Liu, 2019).

Consequently, social identification, trust and value are critical factors in creating long-term relationships. In terms of practical implications, hospitality brands should use social media by tapping into all the opportunities provided by different social media types to create entertaining and trendy messages. For instance, while Facebook can be a suitable medium for longer posts and more information, TikTok can create cliffhangers to draw customers in, and Instagram can be used to generate trendy hashtags and create anticipation. It is also essential that these social media activities are personalized as per customers' preferences and allow interaction to generate word-of-mouth. The information and word-of-mouth generated through these media can motivate customers toward the brand choice (Ebrahim, 2020). This can also facilitate users to identify themselves within the hospitality brand and its social media platforms. These entertaining, trendy, customized, and interactive social media platforms can generate value for users; however, companies should control these activities so that customers' trust towards sharing their experiences on these platforms can be built, leading to long-term relationships with hospitality brands.

Similarly with any other research study, this analysis is also subject to limitations. This study used self-selection sampling and focused on data collection from one platform, i.e., Amazon Mechanical Turk. Moreover, a cross-sectional design was used for this study, which cannot consider the changes in this study's constructs over a time period. Lastly, the study did not see the specific characteristics of several social media platforms and their influence on users' perceptions and behaviors. These limitations may be taken into account in future research. Based on the findings of the study and its limitations, following are some open-ended questions provided for further discussion.

Discussion questions

1. Will the social media activities differ based on the type of social media platforms being used by brands? If yes, which platform(s) might be more efficient for the social media marketing practices performed by the hospitality brands?
2. Does the type of social media platform(s) impact the success of the social media marketing practices performed by the hospitality brands?
3. Do the customers' generation create a perceptional difference in their self-identification with the brand?
4. What will be the trade-off between the role of data privacy concerns and the social media marketing practices?

References

Aksoy, M.E. (2018) A qualitative study on the reasons for social media addiction, *European Journal of Education Research,* **7** (4), 861-865.

Ali, F., Ciftci, O., Nanu, L., Cobanoglu, C., & Ryu, K. (2021) Response rates in hospitality research: An overview of current practice and suggestions for future research, *Cornell Hospitality Quarterly,* **62** (1), 105-120.

Ali, F., Rasoolimanesh, S.M., Sarstedt, M., Ringle Christian, M., & Ryu, K. (2018) An assessment of the use of partial least squares structural equation modeling (PLS-SEM) in hospitality research, *International Journal of Contemporary Hospitality Management,* **30** (1), 514-538.

Bakker, D. (2018) Conceptualising influencer marketing, *Journal of Emerging Trends in Marketing and Management,* **1** (1), 79-87.

Baldus, B.J., Voorhees, C., & Calantone, R. (2015) Online brand community engagement: Scale development and validation, *Journal of Business Research,* **68** (5), 978–985.

Barnes, N.G. & Mattson, E. (2009) Social media in the 2009 Inc. 500: New tools and new trends, *Journal of New Communication Research,* **4** (2), 70-79.

Berthon, P.R., Pitt, L.F., Plangger, K., & Shapiro, D. (2012) Marketing meets Web 2.0, social media, and creative consumers: Implications for international marketing strategy, *Business Horizons,* **55** (3), 261-271.

Bhattacharya, C.B. & Sen, S., (2003) Consumer-company identification: A framework for understanding consumers' relationships with companies, *Journal of Marketing,* **67** (2), 76–88.

Bonds-Raacke, J. & Raacke, J. (2010) MySpace and Facebook: Identifying dimensions of uses and gratifications for friend networking sites, *Individual Differences Research,* **8** (1), 27–33.

Brailovskaia, J., Schillack, H. & Margraf, J. (2020) Tell me why are you using social media (SM)! Relationship between reasons for use of SM, SM flow, daily stress, depression, anxiety, and addictive SM use – An exploratory investigation of young adults in Germany, *Computers in Human Behavior,* **113** (2020), 106511.

Chaffey, D. (2020) *Global social media research summary August 2020,* SmartInsights.com. https://www.smartinsights.com/social-media-marketing/social-media-strategy/new-global-social-media-research/ [Accessed: n.d.].

Chatterjee, P. (2001) Online reviews: Do consumers use them?, *Advances in Consumer Research,* **28**, 129–133.

Chaudhuri, A. and Holbrook, M. (2001) The chain of effects from brand trust and brand affect to brand performance: The role of brand loyalty, *Journal of Marketing,* **65** (2), 81–93.

Chen, S.C. & Lin, C.P. (2019) Understanding the effect of social media marketing activities: The mediation of social identification, perceived value, and satisfaction, *Technological Forecasting and Social Change,* **140** (2019), 22-32.

Chen, S.C., Chiu, Y.B., Lai, M.T., Wu, C.H., Chen, H.H. & Zhang, S.Y. (2014) An empirical study of users' loyalty on virtual communities: The prospective of fundamental interpersonal relations orientation, *Journal of E-Business,* **16** (3), 309–338.

Chu, S.C., Deng, T., & Cheng, H. (2020) The role of social media advertising in hospitality, tourism and travel: A literature review and research agenda, *International Journal of Contemporary Hospitality Management*, **32** (11), 3419-3438

Di Pietro, L., & Pantano, E. (2012) An empirical investigation of social network influence on consumer purchasing decision: The case of Facebook, *Journal of Direct Data Digital Marketing Practice*, **14**, (1), 18–29.

Ebrahim, R.S. (2020) The role of trust in understanding the impact of social media marketing on brand equity and brand loyalty, *Journal of Relationship Marketing*, **19** (4), 287-308.

Elbedweihy, A., Jayawardhena, C., Elsharnouby, M.H. & Elsharnouby, T.H. (2016) Customer relationship building: The role of brand attractiveness and consumer-brand identification, *Journal of Business Research*, **69** (8), 2901-2910.

eMarketer (2016) *Marketing in Asia-Pacific: Millennials in Vietnam are social, skeptical and seeking role models*. www.emarketer.com/Article/Marketing-Asia-Pacific-Millennials-Vietnam-Social-Skeptical-Seeking-Role-Models/1014010.

Farquhar, J. & Rowley, J. (2006) Relationships and online consumer communities, *Business Process Management Journal*, **12** (2), 162–177.

Fornell, C. & Larcker, D.F. (1981) Evaluating structural equation models with unobservable variables and measurement error, *Journal of Marketing Research*, **18** (1), 39-50.

Gallup Pakistan (2019) *Changing consumers and retail behavior in Pakistan-A survey of urban Pakistanis*, https://gallup.com.pk/changing-consumer-and-retails-behavior-inpakistan-a-study-by-gallup-pakistan/ [Accessed: n.d.].

Godey, B., Manthiou, A., Pederzoli, D., Rokka, J., Aiello, G., Donvito, R., & Singh, R. (2016) Social media marketing efforts of luxury brands: Influence on brand equity and consumer behavior, *Journal of Business Research*, **69** (12), 5833–5841. https://doi:10.1016/j.jbusres.2016.04.181

Hays, S., Page, S.J., & Buhalis, D. (2013) Social media as a destination marketing tool: Its use by national tourism organisations, *Current Issues in Tourism*, **16** (3), 211-239.

He, H., Li, Y., & Harris, L. (2012), Social identity perspective on brand loyalty, *Journal of Business Research*, **65** (5), 648–657.

Hollebeek, L.D. (2018) Individual-level cultural consumer engagement styles: Conceptualization, propositions, and implications, *International Marketing Review,* **35** (1), 42-71.

Hollebeek, L., Srivastava, R.K. & Chen, T. (2016) S-D logic-informed customer engagement: Integrative framework, revised fundamental propositions, and application to CRM, *Journal of the Academy of Marketing Science,* **47** (1), 161-185.

Hollebeek, L., Glynn, M. & Brodie, R. (2014) Consumer brand engagement in social media: Conceptualization, scale development and validation, *Journal of Interactive Marketing,* **28** (2), 149-165.

Hughes, A.L., & Palen, L. (2009) Twitter adoption and use in mass convergence and emergency events, *International Journal of Emergency Management,* **6** (3-4), 248-260.

Islam, J., Rahman, Z. & Hollebeek, L. (2017) Personality factors as predictors of online consumer engagement: An empirical investigation, *Marketing Intelligence and Planning,* **35** (4), 510-528.

Ismail, A. (2017) The influence of perceived social media marketing activities on brand loyalty: The mediation effect of brand and value consciousness, *Asia Pacific Journal of Marketing and Logistics,* **29** (1), 129–144.

Kandampully, J., Zhang T., & Bilgihan, A. (2015) Customer loyalty: A review and future directions with a special focus on the hospitality industry, *International Journal of Contemporary Hospitality Management,* **27** (3), 379-414.

Kaosiri, N.Y., Fiol, L.J.C., Tena, M.A.M., Artola, R.M.R., & Garcia, J.S. (2019) User-generated content sources in social media: A new approach to explore tourist satisfaction, *Journal of Travel Research,* **58** (2), 253-265.

Kaplan, A.M. & Haenlein, M. (2010) Users of the world, unite! The challenges and opportunities of social media, *Business Horizons,* **53** (1), 59–68.

Khadim, R.A., Hanan, M., Arshad, A., Saleem, N., & Khadim, N. (2018) Revisiting antecedents of brand loyalty: Impact of perceived social media communication with brand trust and brand equity as mediators, *Academy of Strategic Management Journal,* **17** (1), 1–14.

Kim, A.J. & Ko, E. (2012), Do social media marketing activities enhance customer equity? An empirical study of luxury fashion brand, *Journal of Business Research,* **65** (10), 1480-1486.

Kwon, O. & Wen, Y. (2010) An empirical study of the factors affecting social network service use, *Computers in Human Behavior*, **26** (2), 254-263.

Lam, S.K., Ahearne, M., Mullins, R., Hayati, B. & Schillewaert, N. (2013) Exploring the dynamics of antecedents to consumer brand identification with a new brand, *Journal of the Academy of Marketing Science*, **41** (2), 234-252.

Liat, C.B., Mansori, S., Chuan, G.C., & Imrie, B.C. (2017) Hotel service recovery and service quality: Influences of corporate image and generational differences in the relationship between customer satisfaction and loyalty, *Journal of Global Marketing*, **30** (1), 42-51.

Liu, C-L., Sinkovics, R.R., Pezderka, N., & Haghirian, P. (2012) Determinants of consumer perceptions toward mobile advertising—A comparison between Japan and Austria, *Journal of Interactive Marketing*, **26** (1), 21-32.

Logan, K., Bright, L.F., & Gangadharbatla, H. (2012) Facebook versus television: Advertising value perceptions among females, *Journal of Research in Interactive Marketing*, **6** (3), 164–179.

Mohammadpour, A., Arbatani, T.R., Gholipour, T.H., Farzianpour, F., & Hosseini, S. (2014) A survey of the effect of social media marketing on online shopping of customers by mediating variables, *Journal of Service Science and Management*, **7** (5), 368–376.

Muniz, A.M.J. & Schau, H.J. (2005) Religiosity in the abandoned Apple Newton brand community, *Journal of Consumer Research*, **31** (4), 737–747.

Muniz, A.M.J. & O'Guinn, T.C. (2001) Brand community, *Journal of Consumer Research*, **27** (4), 412–432.

Pham, H., & Nguyen, T. (2019) The effect of website quality on repurchase intention with the mediation of perceived value: The case study of online travel agencies in Vietnam, *Journal of Global Business Insights*, **4** (1), 78-91.

Prahalad, C.K. & Ramaswamy, V. (2004) Co-creation experiences: The next practice in value creation, *Journal of Interactive Marketing*, **18** (3), 5–14.

PWC (2018) *Whom do consumers really trust?*. www.pwc.com/gx/en/industries/consumer-markets/consumer-insights-survey/consumer-trust.html.

Rather, R.A., & Hollebeek, L.D. (2019)'Exploring and validating social identification and social exchange-based drivers of hospitality customer loyalty, *International Journal of Contemporary Hospitality Management,* **31** (3), 1432-1461.

Rather, R.A. & Sharma, J. (2017) The effects of customer satisfaction and commitment on customer loyalty: Evidence from the hotel industry, *Journal of Hospitality Application and Research,* **12** (2), 41-60.

Sanchez, J., Callarisa, L., Rodriguez, R.M., & Moliner, M.A. (2006) Perceived value of the purchase of a tourism product, *Tourism Management,* **27** (3), 394-409.

Schudson, M. (2013) *Advertising, the Uneasy Persuasion (RLE Advertising): Its Dubious Impact on American society,* Oxfordshire: Routledge.

Seo, E.J., & Park, J.W. (2018) A study on the effects of social media marketing activities on brand equity and customer response in the airline industry, *Journal of Air Transport Management,* **66** (2018), 36-41.

Shareef, M.A., Mukerji, B., Dwivedi, Y.K., Rana, N.P., & Islam, R. (2019) Social media marketing: Comparative effect of advertisement sources, *Journal of Retailing and Consumer Services,* **46** (2019), 58-69.

Shareef, M.A., Archer, N., & Dwivedi, Y.K. (2015)'An empirical investigation of electronic government service quality: From the demand side stakeholder perspective, *Total Quality Management and Business Excellence,* **26** (3–4), 339–354.

Sicilia, M. & Palazon, M. (2008) Brand communities on the internet: A case study of Coca-Cola's Spanish virtual community, *Corporate Communication: An International Journal,* **13** (3), 255–270.

Sirdeshmukh, D., Singh, J., & Sabol, B. (2002)'Consumer trust, value, and loyalty in relational exchanges, *Journal of Marketing,* **66** (1), 15–37.

So, K.K.F., King, C., Sparks, B.A. & Wang, Y. (2013) The influence of customer brand identification on hotel brand evaluation and loyalty development, *International Journal of Hospitality Management,* **34**, 31–41.

Sprott, D., Czellar, S., & Spangenberg, E. (2009) The importance of a general measure of brand engagement on market behavior: Development and validation of a scale, *Journal of Marketing Research,* **46** (1), 92-104

Srinivasan, S.S., Anderson, R., & Ponnavolu, K. (2002) Customer loyalty in e-commerce: An exploration of its antecedents and consequences, *Journal of Retailing,* **78** (1), 41–50.

Statista (2020) *Most popular social networks worldwide as of October 2020, ranked by number of active users*. www.statista.com/statistics/272014/global-social-networks-ranked-by-number-of-users/ [Accessed: n.d.].

Stokburger-Sauer, N., Ratneshwar, S., & Sen, S. (2012) Drivers of consumer-brand identification, *International Journal of Research in Marketing*, **29** (4), 406-418.

Tajfel, H. & Turner, J.C. (1986) The social identity theory of inter group behavior, in Worchel S. and Austin W. (Eds.), *Psychology of Intergroup Relations*, Chicago: Nelson, pp. 276-293.

Tatar, S.B. & Eren-Erdogmus, I. (2016) The effect of social media marketing on brand trust and brand loyalty for hotels, *Information Technology and Tourism*, **16** (3), 249–263.

Tran, G.A. & Strutton, D. (2020) Comparing email and SNS users: Investigating e-servicescape, customer reviews, trust, loyalty and E-WOM, *Journal of Retailing and Consumer Services*, **53** (2020), 101782.

Turner, J.C., Hogg, M. A., Oakes, P.J., Reicher, S.D., & Wetherell, M.S. (1987), *Rediscovering the Social Group: A Self-Categorization Theory*. Oxford: Basil Blackwell.

Tynan, C., McKechnie, S., & Chhuon, C. (2010) Co-creating value for luxury brands, *Journal of Business Research*, **63** (11), 1156–1163.

Verhagen, T., Swen, E., Feldberg, F., & Merikivi, J. (2015) Benefitting from virtual customer environments: An empirical study of customer engagement, *Computers in Human Behavior*, **48** (2015), 340–357.

Verma, V., Sharma, D., & Sheth, J. (2016) Does relationship marketing matter in online retailing? A meta-analytic approach, *Journal of the Academy of Marketing Science*, **44** (2), 206-217

Wang, L., Law, R., Hung, K., & Denizci Guillet, B. (2014) Consumer trust in tourism and hospitality: A review of the literature, *Journal of Hospitality and Tourism Management*, **21** (2014), 1-9.

Wang, X., Yu, C., & Wei, Y. (2012) Social media peer communication and impacts on purchase intention: A consumer socialization framework, *Journal of Interactive Marketing*, **26** (4), 198-208.

We Are Social (2020) *Digital in 2020*, https://wearesocial.com/digital-2020.

We Are Social (2021) *Digital 2021: The latest insights into the state of digital*, https://wearesocial.com/blog/2021/01/digital-2021-the-latest-insights-into-the-state-of-digital [Accessed: n.d.].

Wellman, B. (2001) Computer networks as social networks, *Science,* **293** (5537), 2031–2034.

Xu, X., & Pratt, S. (2018) Social media influencers as endorsers to promote travel destinations: An application of self-congruence theory to the Chinese Generation Y, *Journal of Travel and Tourism Marketing,* **35** (7), 958-972.

Yadav, M., & Rahman, Z. (2018) The influence of social media marketing activities on customer loyalty: A study of e-commerce industry, *Benchmarking: An International Journal,* **25** (9), 3882-3905.

Yadav, M., & Rahman, Z. (2017) Measuring consumer perception of social media marketing activities in e-commerce industry: Scale development and validation, *Telematics and Informatics,* **34** (7), 1294–1307.

Yahia, I., Ben Al-Neama, N., & Kerbache, L. (2018) Investigating the drivers for social commerce in social media platforms: Importance of trust, social support and the platform perceived usage, *Journal of Retailing and Consumer Services,* **41** (1), 11-19.

Zhou, Y., Berezina, K., Hui, W., & Kim, W.G. (2020) Commitment is an act, not a word: Examining members' commitment to travel-related online communities, *International Journal of Hospitality Management,* **91** (2020), 102689.

Zollo, L., Filieri, R., Rialti, R., & Yoon, S. (2020) Unpacking the relationship between social media marketing and brand equity: The mediating role of consumers' benefits and experience, *Journal of Business Research,* **117** (2020), 256-267.

4 The role of social media marketing in the travel planning and decision-making processes of today's travelers

Aysegul Gunduz Songur and Gozde Turktarhan

Introduction

The travel and tourism industry contribute extensively to the modern economic system, as indicated by the *Economic Impact 2019*. According to this report published by the World Travel and Tourism Council (WTTC), the industry represents 10.4% of global GDP, contributing to 10% of global employment (WWTC, 2019). Travel and tourism stand as a set of the most promising sectors in the world while they heavily rely on information, which is indispensable for travelers throughout their travel decision making process. In today's world, where the World Wide Web has become a major information source and marketing communication channel owing to the rapid development of information communication technologies, online marketing of travel and tourism related products and services has replaced offline forms of marketing, which once used to be very largely paper-based. This reality brings with it that consumers are not dependent on travel agencies to plan their next vacations (Wagner & Buhalis, 2013). Also, the evolution achieved in information and communication technologies now allow travelers and tourists to have direct multimedia contact with different aspects of any given destination, such as the culture, heritage, landscapes and different tourist services (Jimenez-Barreto *et al.*, 2020) via travel websites and social media platforms.

Literature review

Given the highly competitive nature of today's travel market, it is no surprise that the inclusion of digital tools to the traditional marketing mix is a prevalent practice for destinations. Amongst these digital tools, social media emerged as the most powerful marketing tool while providing a surprisingly global reach for the destination and tourism organizations at a very low cost. Before delving into why and how social media platforms emerged, the concept of online community should be explored first. The information technology aspect is emphasized in the online community definition of Lee *et al.* (2003); in its simplest definition, it is a cyberspace that allows members to create content; while the definition proposed by Kang *et al.* (2007) focuses on the social aspect, and turns out as a social formation in which people join of their free will and organize the contents by interacting with other members. Online travel communities are defined by Lee by emphasizing the common values of people. According to Lee (2005), communities developed by a group of people who have a common travel taste and desire and adhere to common values, norms and interpretations are defined as online travel communities.

Back in 1968, in a very visionary manner, Andy Warhol said *"In the future, everyone will be world famous for 15 minutes"*, which can now be interpreted as referring to every individual's possibility of attaining fame even for a brief period of time due to the increasing prevalence of social media outlets brought about by the advent of the Internet (Brainyquote, n.d.). To support the same line of thought, Li and Wang (2010) posited that the Internet has evolved into a participatory platform rather than the broadcasting medium it once used to be, and it now enables people to turn into the 'media' themselves for the sake of collaborating and sharing information. Similarly, 'YOU', which TIME Magazine named Person of the Year in 2006, represents all users, (Arola, 2010), which also stresses the individuality aspect and the democratizing potential of social media.

According to Paris *et al.* (2010), the concept of social media is a design that shares information over the world wide web, has content created directly by users, facilitates communication and saves time. The concept of social media is not just a single-channel type of communication. In addition, it contains all types of media created by multiple users, such as travel-related web pages, blogs, and social networking pages. Various social media platforms, the most popular and widely used ones being Facebook, Instagram, Twitter and Youtube, are also viewed as tools for creating online communities comprising of users who share similar interests, activities, and objectives (Bolotaeva & Cata, 2010). Kaplan and Haenlein (2006) defined social media from a broader perspective. According to the researchers, the term 'social media' describes an Internet-based platform and a sharing environment where all users create content. In addition to this, all blogs, social networking sites, virtual gaming environments, and all web pages where users can communicate directly and openly with each other are within the scope of social media.

Generations and travel behaviors

Today's travelers have access to a diverse set of resources for travel related information search thanks to the digitally induced communication tools and platforms. Some other studies in the literature state that today's travelers seek authenticity, want to belong to a group, and question the meaning in their lives, and their motivation to travel is shaped in line with these questions (Nijs & Peters, 2002, Ter Borg, 2003). According to *2021 Travel Trends Report*, travelers will tend to embark on more trips and extend their vacations, use their savings on a dream vacation. *Future Traveller Tribes 2030 Report*, on the other hand, stresses that today's travelers are open-minded and mobile savvy. Besides this, 62% of them use their smartphones before and during their travel, while they take the same number of holidays a year as Generation Z. Also, when it comes to travel inspiration, today's travelers are influenced most by pictures posted by their friends on social media platforms (Amadeus, 2021). Although activities and experiences are high on today's travelers' priority list, having a once-in-a-lifetime experience and traveling to destinations where they can take memorable pictures and have vacations with friends are also important considerations for them, significantly more so than it used to be for previous travelers.

Before delving into generational marketing theories and models and how they are connected to social media marketing in tourism, first, the very phenomenon of generations should be explored. A survey of literature revealed the book named '*The Problem of Generations*' in which Karl Mannheim treated generations as a special type of social class. This study is considered as a seminal study with respect to generations, and emerged from systematic and comprehensive studies based on scientific methods (Bengston *et al.*, 1974: 4; Jaeger, 1985: 275-278). According to Mannheim, individuals coming from the same generation share a common position in the social and historical process, just like individuals who come from the same social class. He also stated that individuals from the similar age groups in a certain society and period express that they have similar feelings and thinking structures, similar experiences and similar behavioral patterns. However, age group is not the only criterion that determines the generation. The society, time and social conditions in which individuals live also affect the generation, together with the age group (Mannheim, 1952: 290-292).

One of the most important generational groupings in the field of social sciences, which divides society into more meaningful and wider groups, is the study conducted by William Strauss and Neil Howe in 1991 (Carlson, 2008). Strauss and Howe, divided society into larger groups consisting of common denominators by taking the historical movements that have important effects on the social level as their reference point and defined the generations that has been discussed quite frequently in the related literature (Strauss & Howe, 1991).

The generations referred to in the marketing literature today are based upon the explanation and definition proposed for generations in the context of American society. However, with the phenomenon of globalization and developments in technology, regional and national differences are disappearing, and new generations become more like one another day by day (Levitt, 2006). In the literature, many generations both living today and belonging to historical periods have been described. However, there are differences of opinion regarding the year intervals covered by the generations. The widely accepted generational classification as per marketing theory are Baby Boomers and the generations X, Y and Z. The Baby Boomer generation covers those born between 1946-1964 (Green, 2006: 6; Moschis *et al.*, 2000: 4; Harpaz & Snir, 2015: 255), Generation X comprises of individuals born between 1965-1979 (Erickson, 2010; Green, 2006; Harpaz & Snir, 2015), Generation Y represents the people born between 1980-1994 (Hor & Keats, 2008; Lamanna *et al.*, 2014; Harpaz & Snir, 2015) and Z generation covers those born between 1995- 2009 (Hor & Keats, 2008; Symeonidis *et al.*, 2009; McCrindle & Wolfinger, 2011).

The name Baby Boomers denoting the generation of individuals born between 1946 and 1964, was coined due to the significant number of babies born during the 19 years' period it covers. In the USA, 75.8 million babies were born in this 19-year period, and this figure corresponds to 40% of the population of the USA in 1964 (Green, 2006). For this particular generation, names such as *The Sixties Generation, Flower Children, The Me Generation,* and *Yuppies* are also used (Cochran *et al.*, 2009). Baby Boomers are the first generation to grow up with television. Therefore, television has a significant impact on the character shaping of the Baby Boomer generation (Ball & Gotsill, 2011: 18). In addition to television, especially in the '60s, there have been other important events and milestones that shaped the basic values and characteristics of the Baby Boomer generation, such as the Vietnam War, the assassination of the US President John Fitzgerald Kennedy, the assassinations of Robert Francis Kennedy and Martin Luther King, the Cold War, the anti-racist civil rights movement, the women's movements, the sexual revolution, the environmentalist movements, the anti-war protests, the moon landing, the rock music trend and the widespread use of drugs etc. (Green, 2006; Cochran *et al.*, 2009; Bova & Phelan, 2011; Erdheim & Lodato, 2013;).

Generation X is the generation that includes individuals born between 1965-1979, and is also referred to as Baby Busters, due to the fact that high birth rates seen in the Baby Boomer generation declined between the years covered by this generation (Celek *et al.*, 1996; Hawkings & Mothersbaugh, 2010; Mello, 2015; Mitchell, 1995; Nelson & Quick, 2013). In addition to the negative events that have an impact on the shaping of the characteristics defining the X generation, positive developments regarding technology have influenced the formation of their characteristics. During this period, the first personal computers appeared, and video games became popular amongst the young people. The emergence of the Walkman (portable tape player), Discman (portable CD player) and the start of broadcasting of the MTV music channel coincide with the youth period of the X generation and are among the significant phenomena frequently emphasized when defining this generation (Erickson, 2010).

Generation Y is the generation that includes individuals who were born between 1980-1994. Similar to the baby boomer generation; Generation Y is also defined by an increase in birth rates. Generation Y has the distinction of being the most populous generation after the Baby Boomer generation. Because of their demographic similarities with the Baby Boomer Generation, they are also called as Echo Boomers and Boomlets. As for the generation Y; names such as Generation Next, Generation Why, Millenials, Net Generation, Dot.com Generation, The Connected Generation etc. are also used (Ball & Gotsill, 2011; Sheahan, 2010; McCrindle & Wolfinger, 2011). Generation Y individuals had a more comfortable childhood compared to the generations before them, as they were born at a time when the importance of children for their respective families and society had increased (Strauss & Howe, 1997). Generation Y individuals have witnessed many changes in popular culture. The era of multi-channel television, one of the most important factors affecting the change in popular culture, coincides with the first years of the Generation Y. These new television programs, which usually feature a Western lifestyle, have become accessible to almost everyone from many parts of the world. The Western lifestyle displayed in television programs has been one of the most important factors shaping the expectations and desires of Generation Y individuals (Parment, 2012). Generation Y individuals comprise the very first generation to grow up with the digitally enhanced communication tools triggered by the Internet, such as mobile phones, tablets, etc. as well as the era of multi-channel television. Having been exposed to novel technologies ever since their birth, technology has become indispensable for the life of the Millennials. This close relationship between the Millennials and the Internet and digital technologies is also reflected in their behavior (Ball & Gotsill, 2011).

Generation Z covers individuals born between 1995 and 2010. The year of 1995, marking the starting point for this particular generation, coincides with the commercial worldwide spread of the Internet access (Horn, 2013: 2). For the Z generation, several different names such as Gen Z, Zeds, IGen (internet generation), Digital Natives are also used (Ericsson Consumer Lab, 2012; Grail Research, 2011). Generation Z individuals, through their childhood and early youth, witnessed negative developments such as global crisis, environmental problems and terrorist incidents and were directly or indirectly affected by these developments. In addition to these negative developments, they also witnessed the development and widespread use of digital technology tools based on the Internet and social media, and electronic tools such as laptops, tablets, and cellphones (Grail Research, 2011). Also, Generation Z is the first global generation, and they have had access to almost standard and similar opportunities and services in a global sense, thanks to globalization. Generation Z individuals are found to be emotionally attached to technology and virtual environments; digitally enhanced communication and social media comprise almost the most critical components of their personal life. They seem to be, in a way, addicted to digital communication technologies and social media and it is not easy for Generation Z to live in an environment with out socializing in digital world. For example, according to a study conducted in the USA, stress symptoms were observed in 79% of children who are not allowed to use computers and mobile phones (Grail Research, 2011).

Due to being exposed to and/or witnessing negative events such as war, terrorist incidents, social events and environmental problems, Generation Z individuals have turned into more cautious and pragmatic individuals, while also being willing to take responsibility for changing the world (Sladek & Grabinger, 2013). Individuals of this generation, who are in a constant state of interaction with each other, are raised as the "citizens of the world" (Sladek & Grabinger, 2013). Via social media platforms, they can share all the positive and negative developments in their own environments with the rest of the world, with many types of audio-visual content such as videos, photos, texts, etc. and just the same way, they can stay current about what is happening with the rest of the world. The fact that the internet and social media carry communication and interaction to a whole new level, namely to a global dimension, presents a tremendous opportunity for younger generations of the world to be in constant communication with each other, which would also enable them to make some common decisions for the future of the world. When marketing practices are shaped based on the generational theory, it is understood that the consumption habits of baby boomers and Gen X requires one-way marketing (Fennell, 1982). On the other hand, Gen X has tried to bring a social perspective to marketing and marketers, thus, mass marketing practices have started to become diversified and customized (Dou *et al.*, 2006).

The market domination has been transferred to the consumers with the participation of the Generation Y. However, due to the postmodernist paradigm which opposes the dominance of the mind and the only truth, the Gen Ys have turned to different, private, and personal consumption practices and habits (Noble *et al.*, 2009). In this case, it is possible to say that the expression of differences and the globalization of an unlimited consumption phenomenon are stressed (Parment, 2013). The fact that the Generation Y individuals follow all kinds of digital information and social media platforms has brought marketers face to face with an active consumer base who know about their rights and can immediately ask for an account of whenever necessary. Along the same line, business practices lacking dishonesty, practicing non-transparent and non-accountable business endeavors are immediately posted on social media and taken to the blacklist (Lissitsa *et al.*, 2016). Gen Z, on the other hand, from the moment they start to go for shopping with their parents, constitutes the consumer generation of today and the future, with which marketers try to establish relationships, wish to place their brands into their subconscious and thus shape them accordingly (Posnick-Goodwin, 2010).

Each generation has its own characteristics and generations progress by norms in society (Pendergast, 2010). According to Glover (2010), this whole knowledge about generations presents a common ground for tourism and travel marketers to learn about and compare and contrast the behavior of travelers and tourists based on their generation. By doing so, tourism marketers can gain insight into the consumption patterns of generations and structure their marketing strategies accordingly. For example, destination marketers can segment tourism markets, understand their needs and desires, develop new products and services, and implement marketing strategies using generational differences (Huang & Lu, 2017). For example, Baby Boomers are easily influenced by mass marketing and communication practices, while Generation Ys are affected by interactive, participatory marketing efforts containing networking, viral marketing, and digital communication (Pendergast, 2010).

According to Virtuoso (2018), Generation X spends slightly more on holidays than other generations, and 82% of them prefer online booking. 71% of Generation X tourists were observed to use online travel agents, reviews, and online apps to make purchasing decisions for their vacation (Vivion, 2016). Ian Yeoman (2013) stated in Tedx Talks that the future tourist will be more demanding, more complex, more sophisticated, and will want something at a cheaper price; from a business perspective, this is about understanding the consumer and trying to reinvent themselves.

Pennington-Gray and Blair (2010) posited that Generation Y is the most adventurous generation ever. Along the same line of thought, Pendergast, (2010) stated that Generation Y does not escape from risks or problems such as terrorism and natural disasters, but instead, they tend to reduce risks through information gathering. According to a report published by Megatrends in 2017, tourists are expected to be brave and less fearful in 2030 than previous generations (Megatrends, 2017). Skinner *et al.* (2018), compared Generation Z and Generation Y tourists and found out that they were similar in the sense that they love adventure and gamification experiences that involve realizing physical challenges. Again, according to the Megatrends report, Generation Z tourists would not prefer package holidays, but instead, they would seek unique holiday experiences just like Generation Y tourists (Megatrends, 2017). In addition, Generation Y has the lowest income as they are still new to work life, so this generation's travel budget is limited compared to other generations (Li *et al.*, 2013).

Y and Z generations are also known as digital nomads because they use digital devices as a part of their daily lives. For this reason, their travel behaviors and information sources about their travels are different from other generations. Being away from their homes longer, traveling longer, participating in more activities and gaining more experience are other characteristics that distinguish them from other generations. It is known that the Z generation makes more online reservations than the Y generation, they are more social, and they communicate more with the people living in the region in the destinations they travel to (Vivion, 2016).

According to the results of a study conducted by Booking.com, 60% of travelers using Booking.com in 2018 use and share social media accounts almost every day. This sharing rate varies depending on the age, and younger generations share more frequently. Similarly, Generation Y uses social media quite frequently during their holidays and create content using Facebook, Instagram and Tripadvisor platforms. In addition, according to another study, 27% of millennials living and traveling in the United States review posts on social media platforms before booking online (Expedia and the Center for Generational Kinetics, 2018). After making an online reservation, they continue to review the content created by users on social media platforms (Dedić, 2017).

Social media platforms

Parallel to the emergence of the Internet, the proliferation of social media platforms presents a practical and effective ground for use by several sectors, including tourism, to carry out online marketing activities. Tourism is among the sectors which quickly adopt technological innovations due to the growing need for mutual communication and interaction, and social media marketing has proved to offer powerful yet low-cost marketing tools which can ensure a global reach for travel and tourism enterprises.

Social media enables tourism enterprises to share service and product promotions, marketing activities, announcements, and various campaigns with a wider audience. The dynamic structure of the tourism industry, where production and consumption are almost simultaneous, necessitates the wide use of social media as a marketing tool. The participation of the tourism industry in social media is important for a more effective development with respect to social, economic and cultural dimensions. Therefore, getting involved in social media marketing is a must for the tourism industry. In this respect, according to Zerfass, *et al*, (2009: 60), social media is important due to the following factors:

- Social media is innovative and understandable,
- It is consumer oriented,
- It encourages new ideas and starts communication,

- It allows the establishment of new relationships and partnerships,
- It plays an important role in running viral campaigns.

Social media along with the evolving tourism culture, changes the travel planning process of tourism consumers and travelers and the way they obtain information about tourism destinations. Therefore, social media marketing, in the context of the tourism industry, brings about a new dimension to management flow, such as collecting consumer information, recognition, creating brand awareness and new communication channels (Zeng & Gerritsen, 2014: 33).

Today, tourism consumers see social media as an important information channel when deciding on hotel choices. Social media in this respect offers a free environment for tourism consumers to express themselves through interaction, cooperation, sharing and naturally flowing communication. Tourism enterprises, which attach importance to mutual communication, significantly benefits from being on social media platforms and web sites (such as Instagram, Facebook, Twitter, Tripadvisor, etc.) to increase interaction and accessibility.

According to Dwityas and Briandana (2017: 196), the popularity of social media platforms by tourism consumers during the decision-making process regarding tourism products can be explained in three stages: pre travel, travel, and post-travel stage.

- *Pre-travel stage*: (1) Formation of request and request for travel. (2) Collection and evaluation of information about the destination to be traveled and tourism activities. (3) Booking flight tickets and hotel, which are necessary to make the travel decision and to realize the trip. It includes the pre-purchase of products.
- *Travel stage*: This is the stage where tourism consumers get their travel experience. It includes the consumption of basic needs within the scope of tourism such as accommodation, transportation and food. At the same time, researching other additional information needed within the framework of tourism products is also included in this phase.
- *Post-travel stage*: It is the stage in which the travel activities are carried out, in short, it is the stage where the trip ends. This process includes the holistic evaluation of experiences, tourism products, travel decisions and information sources. The evaluation result is important as they also influence the future travel decisions depending on whether the tourism consumers are satisfied or not.

Tourism consumers can directly connect with many people and share their holiday experiences thanks to social media enabling effective and fast communication. In that respect, social media allow the tourism industry to have an important role in delivering promotional and marketing information to the masses. Social media platforms also contribute to the dreaming stage for tourists, as they provide unlimited information for tourism consumers before their travel starts, and help them to get ideas to envision the hotel they want to stay in (Rathore *et al.*, 2017: 752). Therefore, the rapid adoption of social media by tourism consumers is an indication that the tourism industry should not remain indifferent to social media marketing.

Social media use and E-WOM in travel and tourism

According to Kang (2011), for many travel companies, web platforms where users can create content can be a marketing tool with high returns for the companies. According to the researcher, travel companies can achieve this return in several ways. The first of these is to obtain feedback by analyzing and examining the content created by users in the online environment. The second is to determine the pulse of the market in short time intervals and to generate additional products, promotions and offers according to the needs. The third and last one is to instantly observe whether the services and products they offer provide consumer satisfaction.

Consumer-initiated (created voluntarily by consumers) and company-initiated (supported and produced by a firm) online brand communities are the most common. It is known that the information on web platforms where users create content is a result and reflection of the experiences of users and consumers. From the consumers' perspective, this leads to a certain level of anticipation regarding the product and service being discussed, similar to how the information was provided. When the anticipation is satisfied and the information regarding the product and service is deemed to be trustworthy, then consumers develop a certain level of trust in the brand, which would enhance the psychological attachment to both the online community providing the trustworthy information and the company supplying that particular product and/or service offering. Online communities enable companies to build trust and commitment toward their companies, brands, products and services (Ulusu, 2010) based on the bilateral communication among the online community members regarding one another's consumption experience with the brands. From the company perspective, through the formation of an effective online community, strong trust and commitment toward certain brands can be fostered. For the corporate marketing practices to be successful, managing online communities in an optimal fashion turns out to be highly significant.

According to Huang *et al.*, (2010), social media channels are divided into four categories based on their purpose:

1. The first provides electronic access to shared data;
2. The second provides a framework for consumers to exchange electronic information;
3. The third allows individuals to create personal profiles;
4. The fourth enables members to develop their own pages.

Thus, it should not come as a surprise that the outcomes of social media sharing such as blogs, social media tagging, social networks, along with photographs and videos shared on social media platforms and online travel agency (OTA) websites like Trip Advisor etc. have become increasingly popular amongst modern travelers. Torres (2010) stated that 84% of leisure travelers use the Internet as a resource for planning their travels as per Google research, with social media content constituting almost two-tenth of the search results (Xiang & Gretzel, 2010).

According to Litvin *et al.* (2008), UGC and eWOM represent the same concept and are improved versions of WOM, focusing on the needs of the new generation. Although WOM and eWOM are parallel, eWOM is not restricted by the size of any one social network (Hart & Blackshaw, 2006); instead, it transcends sociocultural barriers due to its omnipresent origins, which is independent of geographical boundaries. The audio-visual content created and shared in the form of texts, pictures, videos, blog posts etc. is called as user generated content (UGC) and constitute eWOM in the background of the social media platforms and social media environment.

There is a huge body of literature suggesting that WOM constitutes a very powerful and influential marketing tool and marketers discovered the importance of WOM, especially in terms of triggering trust and loyalty factors (Soteriades & Van Zyl, 2013). This especially holds true in the context of tourism product and service offerings, due to their intangible nature necessitating interpersonal communications to become an important information source amongst tourism consumers (Litvin *et al.*, 2008). In the new world of interactive marketing where the institutions' touch points with their target groups have so widely dispersed via various forms of media, developed ever since the emergence of the Internet, conventional WOM has been largely replaced by the ubiquitous eWOM. Due to the experiential and intangible nature of hospitality products, eWOM is of crucial significance for any given tourism and hospitality consumer. Word-of-mouth communication has been identified by Kimmel and Kitchen (2014) as the most potent digital media source.

Gruen *et al.* (2005) posited that eWOM communication has a positive impact on the perceived value of a company's products and the loyalty intentions of their customers. Within the context of online marketing, eWOM denotes any form of visual, aural, and audio-visual UCG which may take the form of online reviews (eReviews) that today's hospitality consumer comes across on social media platforms, online recommendation sites and/or hotel interactive web sites. A brief review of literature revealed that social media platforms are also used on the post-trip phase for sharing travel related experiences (Zivkovic *et al.*, 2014). Tourists may need to adjust their vacation plan in line with information shared on social media sites before making any decisions (Roque & Raposo, 2016). Content from travel agencies and tour companies, as well as other mainstream media marketing platforms, are considered more credible than advice from virtual forums (Fortis *et al.*, 2012). With social media, tourists become the co-producers of an extensive amount of information and knowledge, termed as user-generated content and share them with prospective tourism and hospitality consumers. In any given online communication context, the two poles of communication turn out to be *narrator* (as opposed to the *sender* in the traditional communication context) and (ii) *receiver* (as opposed to the *recipient* in the traditional context) (Soteriades & Avgeli, 2007).

Tourists must collect complete data before deciding to buy tour packages, to lessen the threat connected with the acquisition of tourist industry products and activities, which stem from the innovative features related to many touristic experiences (Roque & Raposo, 2016). Therefore, visitors today rely often on knowledge obtained from one another to minimize the threat and insecurities involved with the purchase of tourism services. Sparks and Browning (2011) suggested that a larger effect is associated with brand choices of the consumers, based on positive or negative views of past consumers, than on data obtained through advertising. Along the same line, it was noted that the data provided by Internet partners affect almost as much as the data supplied by businesses' views, choices, and preferences (Constantinides & Fountain, 2008). It is often thought to be similar to the offline advice gathered from friends and relatives, as well as other tourists, for knowledge sent by other tourists on digital networks. (Koppar, 2015; Fridolf & Arnautovic, 2011; Chan & Guillet, 2011). Previous studies also support the concept that consumer data (WOM) is thought to be more trustworthy than data provided by the suppliers, depending on the trustworthiness of electronic database; customer comments are generally seen to be much more genuine than unrecognized vendor/business information (Wilson & Sherrel, 1993).

Along the same line, Wernerfelt (1994a) proposed that online consumer reviews should be deployed as a new element in the marketing communications mix as they function almost in the form of freely employed 'sales assistants' for any given online seller, with the purpose of helping consumers to identify products that best match their needs. As early as 1995, began to offer Amazon.com their consumers, the option of posting their own comments regarding the products displayed on the website (Chen & Xie, 2008), and today Amazon.com has around 250 million consumer reviews on all its product categories, which are regarded as one of the most successful features of Amazon (Review42, 2021) and these stand as a leading example for several other online sellers.

eWOM is considered to be more influential than traditional WOM mainly thanks to its inherent information technology aspect, which enables the opinions of any single individual to instantly reach millions of consumers regardless of time and space (Buhalis & Law, 2008), while also being easily tracked down by the consumers via various forms of social media, from the most current one to the oldest, all of which can remain online for years to come, mostly in an anonymous fashion (Tham *et al.*, 2013). As opposed to traditional WOM, the impact of eWOM or UGC, mostly in the form of online consumer reviews, can reach far beyond the local community, without restrictions of time and space, as consumers from all corners of the world can access eWOM via social media platforms over the Internet.

The enormous growth in travel-related choices, especially in the past few years, is a basic indicator of how online networking platforms have revolutionized the tourism business (Xiang & Gretzel, 2010). Gros (2012) additionally noted that Internet channels may lead to significant alterations in travel behavior, because visitors generally look for views and advice from other customers when making up their minds regarding which products to purchase. Fortis *et al.* (2012) have posited in another stream of UGC studies, that the reviews from the mainstream press disseminated on social networking sites, especially by prominent people, can always have a significant impact on travel plans and the ultimate decision processing cycle. Shyle and Hysi (2015) also claimed that the ways in which passengers collect the data required for destination selection have changed in digital networking. Today customers rely on other tourists' knowledge to organize their holidays and decrease unpredictability (Hua *et al.*, 2017). Roque and Raposo (2016) have indicated why digital networks have become a significant source of knowledge for tourists and how they are using important instruments which cannot be neglected by the tourism industry.

Why hospitality businesses turn their marketing efforts to social media platforms

Social media platforms, where online communities gather, allow marketers to achieve an advanced degree of customization by being able to thoroughly understand and acknowledge the specific needs of each consumer through monitoring content posted by them. Along the same line, Sigala (2003) posited that social media platforms enable marketers to obtain information about both current and prospective customers based on their profile information; come up with inferences regarding consumers' desires, needs and preferences as per their browsing and social media usage history; and obtain direct and real-time replies and comments from consumers.

The particular reasons why hospitality enterprises turn their marketing efforts toward social media platforms for the purpose of creating more engaging and direct relationships with their audiences and the specific ways how they do this, can be better understood if the stages of service consumption are initially explored and then expanded into the context of tourism and hospitality products. According to Lovelock (2011), service consumption comprises of three stages: the stage prior to the purchase; the stage of service encounter; and the stage following the service encounter. The stage prior to purchase, or prior to travel for tourism consumers, refers to when the purchasing decision is made by evaluating the information received regarding that particular service offering. This considerably complex and time-consuming stage of pre-purchase decision-making requires more time to complete in comparison to the actual purchasing process and it involves various factors (Fisk, 1981). Mattila and Wirtz (2000) suggested that the purchase decision-making for a service offering bears a high degree of uncertainty and customers spend extra time trying to retrieve relevant information in order to minimize their risk. This particular stage of purchase decision-making is exactly where the social media comes into play with a significant role by providing consumers with an indispensable access to the real experiential information from the consumers who have actually consumed the service offering in question.

Within the broad context of hospitality and tourism industry, the fact that the information accessed via social media platforms, in contrast to traditional information resources such as catalogues, brochures, travel guides, magazines, etc. is both low-cost and widely accessible (Gursoy & McCleary, 2004), also stand out as advantages for the tourism consumers. Because of the fact that hospitality and tourism products and service offerings are intrinsically intangible, they possess a higher perceived risk for the prospective consumers, and UGC or eWOM in the form of customer recommendations and blog posts on social media platforms plays a significant role in terms of decreasing this risk. Some people in the industry understand that social networking recommendations (especially online comments) and customer evaluations (accumulated values depending on customer evaluations), have an effect on customer purchasing habits (Noone & McGuire, 2013). In the words of Schmallegger and Carson (2008), the evaluation of the quality and the associated benefits of travel and tourism products is impossible before the product is consumed. In this respect, the UGC or eWOM ubiquitously found on social media serve as additional sources of information for travelers (Xiang & Gretzel, 2010). Sometimes consumers perceive UGC or eWOM less trustworthy than the traditional WOM due to their anonymity, which is a natural outcome related with the difficulty of message source identification (Sparks & Browning, 2011). On the other hand, Burgess *et al.*, (2009) came up with a contradictory suggestion that UGC is considered as trustworthy, as it denotes real experience revealed by real people, and stated further that UGC is considered to be more trustworthy if it comes from a travel related online community rather than coming from a more generic social media platform.

The new marketing mix with social media platforms for tourism and hospitality organizations

The marketing communications environment has been revamped following the emergence of social media, as the dominating digital communications channel. Due to the rapid expansion of social media channels and networking platforms, it is necessary for tourism and destination marketers to treat them as another marketing tool and carefully devise ways to include them in their integrated marketing communication strategy. Various needs of today's consumers are catered for via social media channels, such as making new friends, socializing, content creation, and sharing and exchanging online word-of-mouth comprising of their relevant experience, opinions and information regarding any given product or service offering (Khang *et al.*, 2012). The most significant impact of social media channels as a novel type of communication platform over the service consumption process is mainly due to its inherent feature of allowing communication to be two-way, where it once used to be unidirectional; thus, it enables companies to reach consumers who are online at any given period of time and get the advantage of online word-of-mouth shared between them (Heinrichs *et al.*, 2011). In addition to serving as the most relevant communication channel, allowing for interaction amongst consumers themselves along with interaction between customers and companies, social media platforms also emerge as the most indispensable information source where modern consumers can produce, disseminate and gather information as per their purchasing needs. As an informative venue contributing to the relationship amongst customers through experience sharing, social media also provide valuable information about consumers' needs and wants for the suppliers.

The hospitality industry has quickly adopted this paradigm shift in marketing strategies triggered by social media usage. As opposed to the conventionally known and accepted four principles of marketing (Product, Price, Promotion, Place), the principles of social media marketing are described by Birch (2011) as the four Rs (Relationship, RoE/Return on Engagement, RbR/Reach based on Relevance, and Reputation). Barefoot and Szabo (2010) posited that social media marketing, in its essence, is about building relationships and briefly describe it as making use of social media platforms for the purpose of promoting a company and its products. Social media marketing relies on user engagement and unlike other forms of marketing practices, the consumers are no longer passive receivers of product information, but they also take part in the process and become co-creators of the marketing messages. The logic of marketing in the new age, as posited by Vargo and Lusch (2004), relies on value and relationship co-creation along with intangible resources.

The growing interest in social media platforms in the tourism industry is mainly due to the fact that they comprise powerful yet low-cost marketing tools, with a distinctively immense global reach. The growing popularity of online communication and the development of online networks have revolutionized the way tourist attractions advertise their offerings over the recent years (Tafveez, 2017). Also, the state-based social media networks have become key providers of travel information (Roque & Raposo, 2016). To attract today's visitors, Destination Marketing Organizations and hospitality enterprises have been incorporating digital marketing tools, including social media platforms, into their traditional marketing practices. Digital networks are growing significant for almost every tourism organization and are components of the tourist activities that effect profitability. Along the same line, a creatively thought-out communication mix comprising of various social media channels, facilitates companies' connection with their existing clientele and prospective customers (Chang *et al.*, 2015).

As posited by Buhalis and Law (2008), with the emergence of social media channels, the way information is created and disseminated by the society has evolved, thus both the supply (indicated by marketing/management practices of hospitality and tourism organizations) and demand (reflected on the decision-making process of consumers) aspects of the tourism industry have been altered accordingly. Tourism destinations and businesses have come to understand that the consumption behavior of today's travelers, and/or consumers of tourism and hospitality products has been vastly changed due to information communication technologies and social media usage (Soteriades & Van Zyl, 2013). Thus, the future success of destination marketing organizations and hospitality businesses in developing, marketing and distributing their service and product offerings will be in parallel to the way they predict how to reach these consumers, and social media marketing will continue to play a key role for the industry as a whole.

While Facebook and Twitter used to be the main focus of research with respect to how social media channels are used in travel and tourism industry as a result of their relatively high number of global account ownership and active users, (Golbeck, 2015), Instagram and Youtube have recently appeared as additional social media platforms for tourism researchers, mainly owing to their visual rather than textual content. According to Hays *et al.* (2013), Facebook comprises a multi-faceted platform with both visual and textual content, but Twitter is mostly text-oriented, enabling users to post messages called 'tweets' with maximum 140 characters (Jansen *et al.*, 2009). On the other hand, Youtube, launched in 2005 by Google, as a global video sharing platform, and Instagram, a relatively novel type of social media channel, launched in 2010 by Facebook, enable their users to post their photos and/or short films, with a duration of maximum 60 minutes.

Enabling users to share and have access to an increased volume of visual content makes Instagram and Youtube increasingly favored social media channels for travelers, while at the same time they constitute newly emerged marketing channels for DMOs, and tourism and hospitality establishments. According to Nusair *et al.*, 2008, different personalities of consumers and the values which motivate them to engage in certain types of online behavior can be described as both hedonic and utilitarian in nature.

In that respect, using travel related social networking sites and social media platforms serve both of these values; while seeking information serves the utilitarian value, surfing the Internet and spending time on social media for fun and entertainment purposes serve the hedonic value. Travelers tend to use social media platforms as communication channels to share their travel related information, personal memories and experiences with an underlying motivation shaped by both utilitarian and hedonic values.

Social media platforms contribute to hospitality, travel and tourism enterprises as a valuable managerial tool providing them with substantial amount of strategic information achieved through the collection of consumer feedback which helps to retain the existing clientele and attract new customers based on an effective customer relationship management for the sake of strengthening the loyalty to the brand (Leung *et al.*, 2013). Strategic marketing using innovative marketing tools like social media platforms mostly relate to the level of online presence and selection of which channel(s) to use in terms of creating additional business value along with acquiring new customers and retaining the existing customer base. Knowing that many of the social media platforms and online channels can be linked to each other, like videos posted on Youtube can be inserted by adding a hyperlink to other social media posts, etc., a certain level of online presence synergy can also be achieved. Hamill *et al.* (2012) classified the online presence level for business enterprises by three different scales such as 'High', denoting an online presence in more than half of the existing social media channels, 'Medium', referring to an online existence in half of the channels and 'Low', to refer to an online existence in just a few of them.

The strategic decision making, that marketers need to make at this point involve questions as to: (i) which social media channel to choose; (ii) why to choose them; and (iii) how to communicate within each chosen medium. Each medium selected for the online presence should be relevant to and targeted toward specific marketing goals (Gunelius, 2011). The online presence should be orchestrated well and all the online activities that the company is engaged in should be in perfect alignment across all channels (Kaplan & Haenlein, 2010). The strategic information obtained from social media platforms serves the various additional management functions of the enterprise (Litvin *et al.*, 2008); such as Research and development (product improvement, creation of new product and service offerings), product distribution, services marketing (discovering the visitors' experience and solving visitor problems during both pre-, post- and during the service encounter stage) and marketing communication (advertising, public relations, promotion and reputation management). As such, social media marketing functions encompass every component included in the marketing mix (Yoo & Gretzel, 2010) and their contribution to the marketing mix should not be seen as a replacement, but rather as complementary extensions, of the components comprising the marketing mix (Evans, 2008). Reputation management, which once used to be the task of the Public Relations department for any given business, now seems to be almost entirely tackled by the implementation of the right social media channel selection strategy. Owing to the intrinsically experiential aspect of their service and product offerings, this especially holds true for tourism and hospitality enterprises, which heavily rely on consumer generated content consisting of reviews and opinions of past consumers to secure future business with prospective consumers.

The specific focus achieved by the use of the right social media strategy also enables companies to develop customer relationships in a new sense, thus, to pursue a broadened and extended relationship marketing. Sigala (2012) describes this process from the perspective of tourism and hospitality enterprises, which are now able to have a continuous interaction with the customers that they deem valuable to interact with, and how these customers are integrated into the value creation processes by providing inputs via social media. The principles of relationship marketing as determined by Diller (2000), in the form of 6 Is, which are: (i) Information about customers, (ii) Investment in customers, (iii) Individual standing for customers, (iv) Interactivity with customers, (v) Integrating the customers and (vi) Intending to create a unique relationship. All are achieved in the process of marketing via social media channels but in a fundamentally different form. That is to say, it is now possible to create and maintain a unique relationship with a company's customer base through digital interaction catered to the individual needs of each and every customer, by means of the data available to companies, far beyond than the demographic and transactional data which used to be the norm of the past. Today, the enormous amount of data that can be gathered with respect to customers even allow companies to do behavioral targeting and engage in distinctive segmentation approaches. To encourage tribal marketing, customer to customer interactions are also encouraged via social media marketing. On the other hand, customers can also easily create the very same link with the competitors and it is possible to track down their relationship with the competitors via customers' and/or competitors' social media feeds.

UGC posted on social media platforms helps the enterprises to enhance visitor satisfaction by either improving the existing product and service offerings or adding new ones to their portfolio as per the needs and wants of the customers as articulated on UGC. It also helps with enhancing all three stages related with the travel planning process of the traveler or along the same line, the pre, during and post- phases of the service process. Social media channels play a significant role in the research process of any given hospitality service, including that of travel planning (Yoo & Gretzel, 2010) mainly because the quality of hospitality, travel and tourism products and services cannot be assessed before consumption (Schmallager & Carson, 2008). Researching a hospitality product on social media and coming across first-hand experiential information posted by the previous customers triggers trust in prospective tourism and hospitality consumers toward their purchase decision making process and they finalize their purchase accordingly. Simply by registering as a member on Facebook, which currently is the largest social media platform, with around 2.60 billion active users, and by creating a business page, travel and tourism enterprises can start pursuing international e-marketing. By doing so, they can end up with a tremendous web-based marketing tool to provide information, promote their product and service offerings and respond on a real time basis to their customers (Kavasana *et al.*, 2010).

Despite the prevailing belief that social media function solely as a channel for marketing and dissemination of information, it also has a potential in direct product distribution and revenue generation. For example, promoting hotel room inventory on TripAdvisor is a product distribution strategy employed by lodging companies to increase their revenue generation via marketing links leading to OTA websites (Leung *et al.*, 2013), which encourages social media users to book hotel rooms with just an additional click. Simply by providing a link on social media channels to a hotel website and/or an online engine dedicated to hotel booking, social media users are encouraged to make direct bookings instead of booking through third-party channels (Noone *et al.*, 2011). In a similar line of thought, Ye *et al.*, (2011) in their seminal study, attempting to fill the existing research gap by means of quantification of the effect of the online user reviews on the number of online sales, revealed how online user reviews positively impacted online room sales and pointed out that the efficient use of traveler reviews can enable a tourism business to expand and improve their services in order to gain a competitive advantage.

The social culture, incorporating democracy, community relations, collaboration within communities, along with the formation of authentic, and transparent relationships (Barefoot & Szabo, 2010), upon which interactions via social media channels are built, needs to be thoroughly understood to be able to come up with and implement successful social media marketing campaigns. Along the same line, advertising and selling as they once used to be in traditional marketing have been replaced by participation, sharing, and collaboration (Kaplan & Haenlein, 2010); solidarity and continuing conversation are all stressed in social media marketing. Authentic, participatory, and continued conversation (communication) after the sale is closed needs to be ensured in the broad social media marketing context.

Cultural sensitivity should also be considered while formulating social media marketing strategies and designing marketing campaigns using social media channels. The availability and usage of social media channels differs between countries and the way they are used might also differ significantly due to cultural differences. The global marketing motto 'Think global, act local' should also come into play while designing social media marketing campaigns for different communities, societies and countries with a special consideration on cultural differences and sensitivities. As to how cultural differences are revealed in social media usage patterns amongst users coming from different countries and cultures, Lee *et al.*, (2009) discovered significant differences between how the US and Korean travelers employ social media channels to share their relevant experiences with regard to their touristic activities; while US travelers were found to be more focused on sharing their own personal experiences, the Korean travelers were observed to be focusing more on giving recommendations rather than sharing their own personal experiences.

Traditionally, conversations and/or WOM have always been fundamental factors for inherently experiential tourism and travel related information search and purchase decision-making for travel and tourism product and service offerings. Experiential information sharing may include, to a large extent, personal and destination photographs, and personal narrations of travel and tourism experience, all of which can contribute to inspiration for travel planning and the future vacations of prospective tourists and travelers. Now that there are social media platforms to document such experiential information and memorabilia, which can be engaging for all wanting to travel, the feeling of solidarity and the sense of community amongst travelers are enhanced. There is also the status symbol effect for tourism products, and they are considered as important elements of identity construction of those who purchase them (Lee *et al.*, 2009). In that respect, the platforms serve as a venue for travelers to disclose their personal travel memories which they consider as an extension of their identity.

Conclusion

Social media provide a comprehensive marketing platform even for tourism and travel enterprises operating on a relatively low marketing budget, by leveling the playing fiel,d based upon immense marketing opportunities with a very low entry cost while at the same time ensuring an immense global reach. Just the same way as the traditional marketing principles are applied to formulate the overall marketing strategy, social media facts need to be thoroughly studied and analyzed to design a relevant social media marketing strategy based on current realities of the marketplace for tourism, travel and hospitality enterprises. Due to constantly emerging new digital technologies and the dynamic social media landscape, the social media marketing mix and overall strategy need continuous adjustments, which require laborious efforts on the part of the marketers. To ensure a successful marketing strategy via social media channels, there exists an urgent need for the adoption of data-driven approaches.

Although UGC and content posted on social media channels in the tourism and travel related products' context, have already been tackled by scholars to a certain extent, it remains as a vast area which needs further review and exploration, particularly in terms of synthesizing the available literature, as the digitally savvy travelers of today tend to rely on UGC posted on social media platforms more than ever before. That knowledge is crucial for the industry and used by advertisers in identifying how travelers acquire knowledge digitally and improving their entire digital network advertising approach. In this chapter, based on the relevant literature, inferences have been made as to how today's travelers use social media in travel planning and how social media triggers informed decision making for tourism and travel related product and service purchases.

Discussion questions

1. What was the impact of the Baby Boomers generation on the travel habits of the era?
2. In what way are Generation Z individuals changing travel for older generations?
3. Do you think social media platforms will continue to influence travel habits of the new generations? Why?

References

Amadeus (2021)www.amadeus.com/en/insights/infographic/travel-trends-2021infographic

Arola, K.L. (2010) The design of Web 2.0: The rise of the template, the fall of design, *Computers and Composition*, **27** (1), 4–14.

Ball, K. & Gotsill, G. (2011) *Surviving the Baby Boomer exodus: Capturing knowledge for gen X and Y employees*. Boston: Cengage Learning Press.

Barefoot, D. & Szabo, J. (2010) *Friends with Benefits: A social media marketing handbook*. San Francisco: No Starch Press.

Bengston, V., Furlong, M. & Laufer, L. (1974) Time, aging and the continuity of social structure: Themes and issues in generational analysis, *Journal of Social Issues*, **30** (2), 1-30.

Birch, D. (2011). Travel and tourism is a mature business: Nothing much changes or does it?, in ENTER 2011 International Conference on Information and Communication Technologies in Tourism, 26 January. Innsbruck, Austria.

Bolotaeva, V. & Cata, T. (2010) Marketing opportunities with social networks, *Journal of Internet Social Networking and Virtual Communities*, **2011** (2011), 1-8.

Bova, B. & Phelan, S. (2011) Generational and gender shifts in academic medicine, In W. F. Rayburn & J. Schulkin, (Eds.), *Changing Landscape of Academic Women's Health Care in the United States*, Dordrecht: Springer, pp. 25-38.

Brainyquote (n.d.) *Andy Warhol Quotes*, https://www.brainyquote.com/quotes/andy_warhol_109905 [Accessed: 12 December 2020].

Buhalis, D. & Law, R. (2008) Progress in information communication technology and tourism management: 20 years on and 10 years after the Internet, *Tourism Management*, **29** (4), 609-623.

Burgess, S., Sellitto, C., Cox, C. & Buultjens, J. (2009) User-generated content (UGC) in tourism: Benefits and concerns of online consumers,*17th European Conference on Information Systems*, 417-429.

Carlson, E. (2008) Who are the Lucky Few?, in A. Carlson (Ed.), in *The Lucky Few*. Dordrecht: Springer, pp. 1-10.

Celek, T., Zander, D. & Kampert, P. (1996) *Inside the Soul of a New Generation: Insights and Strategies for reaching busters*, Michigan: Zondervan Publishing.

Chan, N.L. & Guillet, B. D. (2011) Investigation of social media marketing: How does the hotel industry in Hong Kong perform in marketing on social media websites?, *Journal of Travel and Tourism Marketing*, **28** (4), 345-368.

Chang, Y.T., Yu, H. & Lu, H.P. (2015) Persuasive messages, popularity cohesion, and message diffusion in social media marketing, *Journal of Business Research*, **68** (4), 777-782.

Chen, Y. & Xie, J. (2008). Online consumer review: Word-of-mouth as a new element of marketing communication mix. *Management Science*, **54** (3), 477-491.

Cochran, L.J., Rothschadl, A.M. & Rudick, J.L. (2009). *Leisure Programming for Baby Boomers*, Illinois: Human Kinetics.

Constantinides, E. & Fountain, S. J. (2008) Web 2.0: Conceptual foundations and marketing issues, *Journal of Direct, Data and Digital Marketing Practice*, **9**, 231–244.

Dedić, G. (2017) Online recenzije u turizmu, in Slivar,I. Alerić, D and Stankov, U. (Eds.), *Kupovno ponašanje turista*, Sveučilište J. Dobrile u Puli, pp. 169-184.

Diller, H. (2000) Customer loyalty: Fata morgana or realistic goal?, in Hennig-Thurau, T. & Hansen, U. (Eds.), *Relationship Marketing: Gaining Competitive Advantage through Customer Satisfaction and Customer Retention*. New York: Springer, pp. 29-48.

Dou, W., Wang, G. & Zhou, N. (2006) Generational and regional differences in media consumption patterns of Chinese generation X consumers, *Journal of Advertising*, **35**, 101–110.

Dwityas, N.A. & Briandana, R. (2017) Social media in travel decision making process, *International Journal of Humanities and Social Science*, **7** (7), 193-201.

Erdheim, J. & Lodato, M.A. (2013) Generational differences in older workers and retirement, in M. Wang (Ed.), *The Oxford Handbook of Retirement*, New York: Oxford University Press, pp. 573-587.

Erickson, T.J. (2010) *What's Next, Gen X?: Keeping up, moving ahead, and getting the career you want*, Massachusetts: Harvard Business Press.

Ericsson Consumerlab (2012) Generation Z: Understanding the digital lives of India's young mobile users, *Ericsson Consumer Insight Summary Report*, (Oct), 1-8.

Evans, D. (2008) *Social Media Marketing: An hour a day,* Indianapolis: John Wiley and Sons.

Expedia and the Center for Generational Kinetics (2018): Generations on the move, https://viewfinder.expedia.com/wp-content/uploads/2018/01/Expedia-Generations-on-the-Move.pdf

Fennell, G. (1982) Terms vs. concepts: Market segmentation, brand positioning and other aspects of the academic-practitioner gap, in R.F. Bush and S. D. Hunt (Eds.), *Marketing Theory: Philosophy of science perspectives,* Illinois: Marketing Classics Press, pp. 97–102.

Fisk, R.P. (1981) Toward a consumption/evaluation process model for services, in Donnelly, J.H. & George, W.R. (Eds), *Marketing of Services,* Illinois: American Marketing Association, pp. 191-195.

Fortis J., Buhalis, D. & Rossides, N. (2012) Social media use and impact during the holiday travel planning process, *Information and Communication Technologies in Tourism,* **6** (4), 13- 24.

Fridolf, M. & Arnautovic, A. (2011) *Social Media Marketing: A case study of Saab Automobile AB,* Trollhättan: West University.

Glover, P. (2010) Generation Y's future tourism demand: Some opportunities and challenges, in P. Benckendorff, G. Moscardo, & D. Pendergast (Eds.), *Tourism and Generation Y,* Connecticut: CAB International, pp.155-163.

Golbeck, J. (2015) *Introduction to Social Media Investigation: A hands-on approach,* Massachusetts: Elsevier Inc.

Grail Research (2011) Consumers of tomorrow: Insights and observations about generation Z, *Grail Research Consumers of Tomorrow,* (November), 1-16.

Green, B. (2006) *Marketing to Leading-Edge Baby Boomers: Perceptions, principles, practices, predictions,* New York: Paramount Market.

Gros, N. (2012) *The Influence of Social Media on Consumers during their Purchase Decision-Making Process and the Implications for Marketers,* Dublin: Dublin Business School.

Gruen, T., Osmonbekov, T. & Czaplewski, A. (2005) The impact of customer-to-customer travel-related know-how exchange on customer value and loyalty, *Journal of Business Research,* **59** (4), 449-456.

Gunelius, S. (2011) *Ten Laws of Social Media Marketing,* Entrepreneur.com, http://www.entrepreneur.com/article/218160 [Accessed: 3 January 2021].

Gursoy, D. & McCleary, K.W. (2004) An integrative model of tourists' information search behavior, *Annals of Tourism Research*, **31** (2), 353-373.

Hamill, J., Stevenson, A. & Attard, D. (2012) National DMOs and Web 2.0, in M. Sigala, E. Christou, and Gretzel, U. (Eds.), *Social Media in Travel, Tourism and Hospitality: Theory, Practice and Cases*, Farnharm: Ashgate Publishing, pp. 99-120.

Harpaz, I. & Snir, R. (2014) *Heavy Work Investment: Its nature, sources, outcomes, and future directions*, New York: Routledge.

Hart, C. & Blackshaw, P. (2006) Internet inferno: One customer can take down your company, but you can turn the potential nightmare into a boon, *Marketing Management*, **15** (1), 18–25.

Hawkings, D. & Mothersbaugh, D. (2010) *Consumer Behavior: Building marketing strategy*, New York: McGraw-Hill/Irwin.

Hays, S., Page, S. J. & Buhalis, D. (2013) Social media as a destination marketing tool: Its use by national tourism organizations, *Current Issues in Tourism*, **16** (3), 211–239.

Heinrichs, J.H., Lim, J.S. & Lim, K.S. (2011) Influence of social networking site and user access method on social media evaluation, *Journal of Consumer Behavior*, **10** (1), 347-355.

Hor, J. & Keats, L. (2008) *Finders Keepers: How to attract and retain great employees*, Australia: CCH Australia Limited.

Horn, J. (2013) The youth report, *Strategy*, **24** (4), 12-16.

Hua, L.Y., Ramayah, T., Ping, T.A. & Jun-Hwa, C. (2017) Social media as a tool to help select tourism destinations: The case of Malaysia, *Information Systems Management*, **34** (3), 265-279.

Huang, Q. & Lu, Y. (2017) Generational perspective on consumer behavior: Chinas potential outbound tourist market, *Tourism Management Perspectives*, **24**, 7-15.

Huang, Y., Basu, C. & Hsu, M. (2010) Exploring motivations of travel knowledge sharing on social network sites: An empirical investigation of U.S. college students, *Journal of Hospitality Marketing and Management*, **19** (7), 717-734.

Jaeger, H. (1985) Generations in history: Reflections on a controversial concept, *History and Theory*, **24** (3), 273-292.

Jansen, B. J., Zhang, M., Sobel, K. & Chowdury, A. (2009) Twitter power: Tweets as electronic word of mouth, *Journal of the American Society for Information Science and Technology*, **60** (11), 2169-2188.

Jimenez-Barreto, J., Rubio, N., Campo, S. & Molinillo, S. (2020) Linking the online destination brand experience and brand credibility with tourists' behavioral intentions toward a destination, *Tourism Management*, **79**, 104101.

Kang, I., Lee, K., Lee, S. & Choi, J. (2007) Investigation of online community voluntary behavior using cognitive map. *Computers in Human Behavior*, **23** (1), 111-126.

Kang, J. (2011) *Social media marketing in the hospitality industry: The role of benefits in increasing brand community participation and the impact of participation on consumer trust and commitment toward hotel and restaurant brands*, doctoral dissertation, IA State University, Ames, IA.

Kaplan, A.M. & Haenlein, M. (2010) Users of the world, unite! The challenges and opportunities of social media, *Business Horizon*, **53** (1), 59–68.

Kavasana, M.L., Nausair, K., Teodosic, K. (2010). Online social networking: redefining the human web, *Journal of Hospitality and Tourism Technology*, **1**(1), 68 –82

Khang, H., Ki, E.J. & Ye, L. (2012) Social media research in advertising, communication, marketing and public relations, *Journalism and Mass Communication Quarterly*, **1** (1), 68-82.

Kimmel, A. & Kitchen, P. (2014) Word of mouth and social media, *Journal of Marketing Communications*, **20** (1-2), 2-4.

Koppar, R. (2015) Impact of social media on tourism and hospitality, *J. of Management & Commerce*, **2**(1), 24 – 26.

Lamanna, M.A., Riedmann, A. & Stewart, S. (2014) *Marriages, Families, and Relationships: Making choices in a diverse society*, California: Cengage Learning.

Lee, F.S.L., Vogel, D. & Limayem, M. (2003) Virtual community informatics: A review and research agenda, *Journal of Information Technology Theory and Application*, **5** (1), 47-61.

Lee, H.Y. (2005) *Understanding member identification in the online travel communities and member voluntary behaviors*, doctoral dissertation, Oklahoma State University, USA.

Lee, Y.J., Yoo, K.H. & Gretzel, U. (2009) Social identity formation through blogging: Comparison of U.S. and Korean travel blogs, in *Proceedings of the 14th Annual Graduate Student Research Conference in Hospitality and Tourism*, Las Vegas, USA, 4-6 January.

Levitt, T. (2006) *Ted Levitt on Marketing,* Massachusetts: Harvard Business School Publishing Corporation.

Li, X. & Wang, Y. (2011) Measuring the effectiveness of US official state tourism websites, *Journal of Vacation Marketing,* **17** (4), 287-302.

Li, X., Li, X. R. & Hudson, S. (2013) The application of generational theory to tourism consumer behavior: An American perspective, *Tourism Management,* **37**, 147-164.

Lissitsa, S. & Kol, O. (2016) Generation X vs. Generation Y - A decade of online shopping, *Journal of Retailing and Consumer Services,* **31**, 304–312.

Litvin, S.W., Goldsmith, R.E. & Pan, B. (2008) Electronic word-of-mouth in hospitality and tourism management, *Tourism Management,* **29** (3), 458-468.

Lovelock, C.W. (2011) *Services Marketing: People, technology, strategy,* 7th ed., New Jersey: Prentice Hall, Upper Saddle River.

Mannheim, K. (1970) The problem of generations, *Psychoanalytic Review,* **57** (3), 378-404.

Mattila, A.S. & Wirtz, J. (2000) The role of pre-consumption affects in post-purchase evaluation of services, *Psychology and Marketing,* **17** (7), 587-605.

McCrindle, M. & Wolfinger, E. (2011) *The ABC of XYZ: Understanding the Global Generations,* Australia: University of New South Wales Press Ltd.

Megatrends (2017) The changing hotel, tourist and destination, *Megatrends Travel,* (5), 1-8.

Mello, J.A. (2015) *Strategic Human Resource Management,* Ohio: Cengage Learning.

Mitchell, S. (1995) Busters and boomers: Caught in the middle, in Popken, R., Newsome, A. & Gonzales, L. (Ed.), *Departures: A reader for developing writers,* Massachusetts: Allyn and Bacon, pp. 70-73.

Moschis, G.P., Lee, E., Mathur, A. & Strautman, J. (2000) *The Maturing Marketplace: Buying habits of baby boomers and their parents,* Connecticut: Greenwood Publishing Group.

Nelson, D.L. & Quick, J.C. (2013) *Organizational Behavior: Science, the Real World, and You,* South-Western: Cengage Learning.

Nijs, D. and F. Peters (2002), *Imagineering. Creating worlds of experience.* Amsterdam: Boom.

Noble, S.M., Haytko, D.L. & Phillips, J. (2009) What drives college-age Generation Y consumers?, *Journal of Business Research*, **62**, 617–628.

Noone, B.N., McGuire, K.A. & Rohlfs, K.V. (2011) Social media meets hotel revenue management: Opportunities, issues and unanswered questions, *Journal of Revenue and Pricing Management*, **10** (4), 293–305.

Noone, B.M. & McGuire, K.A. (2013) Effects of price and user-generated content on consumers' pre-purchase evaluations of variably priced services, *Journal of Hospitality and Tourism Research*, **38** (4), 562-581.

Nusair, K., Yoon, H. & Parsa, H.G. (2008) Effect of utilitarian and hedonic motivations on consumer satisfaction with travel websites, *Information Technology and Tourism*, **10** (1), 75-89.

Paris, C. M., Lee, W. & Seery, P. (2010) The role of social media in promoting special events: Acceptance of Facebook events, *Information and Communication Technologies in Tourism*, **14**, 531-541.

Parment, A. (2012) *Generation Y in Consumer and Labor Markets,* New York: Routledge.

Parment, A. (2013) Generation Y vs. Baby Boomers: Shopping behavior, buyer involvement and implications for retailing, *Journal of Retailing and Consumer Services*, **20**, 189–199.

Pendergast, D. (2010) Getting to know the Y Generation, in P. Benckendorff, G. Moscardo, & D. Pendergast (Eds.), *Tourism and Generation Y*, Cambridge: CAB International, pp.1-15.

Pennington-Gray, L. & Blair, S. (2010) Nature-based tourism in North America: Is Generation Y the major cause of increased participation?, in P. Benckendorff, G. Moscardo, & D. Pendergast (Eds.), *Tourism and Generation Y*, Cambridge: CAB International, pp. 73-84.

Posnick-Goodwin, S. (2010) Meet generation Z, *California Educator*, **14**, 8–18.

Rathore, A. K., Joshi, U. & Ilavarasan, P. (2017) Social media usage for tourism: A Case of Rajasthan tourism, *Procedia Computer Science*, **122**, 751-758.

Review42 (2021) *What percentage of Amazon reviews are fake?*, https://review42.com/resources/what-percentage-of-amazon-reviews-are-fake/.

Roque, V. & Raposo, R. (2016) Social media as a communication and marketing tool in tourism: An analysis of online activities from international key player DMO, *Anatolia: An International Journal of Tourism and Hospitality Research*, **27** (1), 58-70.

Schmallegger, D. & Carson, D. (2008) Blogs in tourism: Changing approaches to information exchange, *Journal of Vacation Marketing*, **14** (2), 99–110.

Sheahan, P. (2010) *Generation Y: Thriving (and Surviving) with Generation Y at Work*, Prahran: Hardie Grant Books.

Shyle, I. & Hysi, V. (2015) Social media and its impact on decision making for trip, *European Journal of Interdisciplinary Studies*, **1** (1), 8-15.

Sigala, M. (2003) Developing and benchmarking Internet marketing strategies in the hotel sector in Greece, *Journal of Hospitality and Tourism Research*, **27** (4), 375-401.

Sigala, M. (2012). Web 2.0 and customer involvement in new service development: A framework, cases and implications in tourism. In: E. Christou & M. Sigala (Eds.) *Social Media in Travel, Tourism and Hospitality*. Routledge: London, pp. 25-38.

Skinner, H., Sarpong, D. & White, G. R. (2018) Meeting the needs of the Millennials and Generation Z: Gamification in tourism through geocaching, *Journal of Tourism Futures*, **4** (1).

Sladek, S. & Grabinger, A. (2013) Gen Z: The first generation of the 21st century has arrived!, *XYZ University*, 1-15.

Soteriades, M. & Avgeli, V. (2007) Promoting tourism destinations: A strategic marketing approach, *Tourism: An International Interdisciplinary Journal*, **55** (3), 335-345.

Soteriades, M. & Van Zyl, C. (2013) Electronic world-of-mouth and online reviews in tourism services: The use of twitter by tourists, *Journal of Electronic Commerce Research*, **13** (1), 103-124.

Sparks, B. A. & Browning, V. (2011) The impact of online reviews on hotel booking intentions and perception of trust, *Tourism Management*, **32**(6), 1310–1323.

Strauss, W. & Howe, N. (1991) *Generation Z*, New York: William Morrow.

Symeonidis, P., Ntempos, D. & Manolopoulos, Y. (2014) *Recommender Systems for Location-based Social Networks*, New York: Springer.

Tafveez, M. (2017) Role of social media in tourism: A literature review, *International Journal for Research Science and Engineering Technology*, **5** (6), 633-635.

Ter Borg, M. (2003). *De Zineconomie. De samenleving van de overtreffende trap*. Netherlands: Scriptum.

Tham, A., Croy, G. & Mair, J. (2013) Social media in destination choice: Distinctive electronic word-of-mouth dimensions, *Journal of Travel and Tourism Marketing,* **30** (1-2), 144-155.

Torres, R. (2010) Today's traveler online: 5 consumer trends to guide your marketing strategy, *Eye for Travel, Travel Distribution Summit,* **14.**

Ulusu, Y. (2010) Determinant factors of time spent on Facebook: Brand community engagement and usage types, *Journal of Yasar University,* **18** (5), 2949-2957.

Vargo, S. L. & Lusch, R. F. (2004) Evolving to a new dominant logic for marketing, *Journal of Marketing,* **68** (January), 1-17.

Virtuoso (2018) *The truth about how different generations travel,* http:// blog.virtuoso.com/tips-and-trends/the-truth-abouthow-different-generations-travel/ .

Vivion, N. (2016) *What stats reveal about travel behaviors by generation,* Sabre, https://www.sabre.com/insights/whatstats-reveal-about-travel-behaviors-by-generation/ [Accessed: n.d.].

Wagner, R. & Buhalis, D. (2013) E-destinations: Global best practice in tourism technologies and applications, in L. Cantoni and Z. Xiang (Eds.), *Information and Communication Technologies in Tourism,* Springer, pp. 119-130.

Wernerfelt, B. (1994) On the function of sales assistance, *Marketing Science,* **13** (1), 68-82.

Wilson, E. J. & Sherrell, D.L. (1993) Sources effects in communication and persuasion research: A meta-analysis of effect size, *Journal of Academic Marketing Science,* **21** (Spring), 101-112.

WWTC, (2019). Economic Impact Reports. https://wttc.org/Research/ Economic-Impact. [Accessed December 20, 2021]

Xiang, Z. & Gretzel, U. (2010) Role of social media in online travel information search, *Tourism Management,* **31** (2), 179–188.

Yeoman, I. (2013). The Future of Tourism: Ian Yeoman at TEDxGroningen. https://www.youtube.com/watch?v=IRv1-6Y9N2A&ab_ channel=TEDxTalks. [Accessed November 03, 2021]

Ye, Q., Law, R., Gu, B. & Chen, W. (2011) The influence of user-generated content on traveler behavior: An empirical investigation on the effects of e-word-of-mouth to hotel online bookings, *Computers in Human Behavior,* **27**, 634-639.

Yoo, K. H. & Gretzel, U. (2010) Antecedents and impacts of trust in travel-related consumer-generated media, *Information Technology and Tourism*, **12** (2), 139–152.

Zeng, B. & Gerritsen, R. (2014) What do we know about social media in tourism? A review, *Tourism Management Perspectives,* **3** (10), 27-36.

Zerfass, A., Moreno, A., Tench, R., Vercic, D. & Verhoeven, P. (2009) European communication monitor, *Trends in Communication Management and Public Relations*, EACD.

Zivkovic, R., Gajic, J. & Brdar, I. (2014) The impact of social media on tourism, *Singidunum Journal of Applied Sciences Supplement*, **2014**, 758- 761.

5 The Instagram effect on the transformation of tourist behavior to overtourism

Ilke Basarangil

Introduction

Tourists can easily access the information through the Internet; for example, tourists can get information about a destination, provided by people who have experience of it, and decide from different opinions. The rapid advance in technology related to Internet and social networks has led to changes in the demands and needs of tourists, and the emergence of different sales and marketing methods. With the widespread use of Internet, the global communication network, and with the development of social networks, tourist behavior has been rapidly changing. The use of social networks like Instagram has made them the focus of tourism marketers who want to increase profits. Among the behaviors of the new generation of tourist, feedback such as liking, being different, being noticed and appreciated from Instagram have been extremely important. These kinds of feedback affect people quite a lot.

This effect is called the 'Instagram effect', 'bandwagon effect' (herd mentality) and the 'snob effect'. Instagram, which is the most popular social network of recent years, has attracted the attention of social media marketers because of this effect on its users. The rise of Instagram has brought some problems in tourism. Due to the Instagram effect, some problems have begun to be experienced in destinations. The most important of these is overtourism, with its harmful imacts on people and the environment. Overtourism worsens tourist experiences in tourist destinations and lowers the quality of life of local residents. The Instagram effect continues to inspire tourists today about their travel decisions and destination experience. In this research, various examples will be given about how today's tourist behaviors transform into overtourism with the Instagram effect, bandwagon effect, snob Effect. This study is a literature review. As a result of this study, useful suggestions were developed for tourism marketers for the problems of destinations exposed to overtourism due to the effect of Instagram.

"A clear definition of overtourism does not yet exist. By its very nature, the phenomenon of overtourism is related to the number of tourists, the type and time frame of their visits, and the carrying capacity of the destination. Perspectives on overtourism may include the views of various stakeholders, such as residents, tourists or businesses" (Peeters *et al.*, 2018: 21). A great photo should be the result of a great experience, not the reason for going to a destination. But today the concept of travel is changing. People want more than life-enriching experiences or searching for casual destination experiences through social media. The desire to capture the same iconic square as your favorite celebrity contributes to overtourism, which makes the trip less enjoyable for travelers as well as for local residents. (Wayfairer, 2019). Social media applications are the area where companies reach millions of consumers and sell their products and talk about services, not only in travel and travel areas but also in boutiques or businesses marketing through Instagram (Onurlubaş & Öztürk, 2018: 989). Social media also has benefits. The holiday experience of the tourist who can use his social media account at any time, at any place and at any quality and speed is positively affected and any complaints about the internet services provided by the enterprise are minimized (İlban & Güleç, 2018: 71).

Many people travel only to capture a particular image and verify their online presence. If a traveller does not receive hundreds of likes on a page, the traveler may ask if it was worth going. If a person can not post anything about a holiday on social media, would the person go anyway? *"Almost everywhere I travel, I witness the growing trend of "doing it for the 'gram." In every hotspot, you can feel the desperation as people clamour to get their flawless shot. Glamorous women pose at the top of a mountain, changing into gorgeous dresses and reapplying their makeup after a hike. Tourists prance and pose in picturesque locations, posting their images with narcissistic hashtags like #mylifeisbetterthanyours or #fromwhereyoudratherbe, perpetuating a competitive atmosphere."*(Wayfairer, 2019). In this study, the instances of tourist destinations exposed to overtourism by the Instagram effect and the behavior of tourists causing overtourism have been examined in detail. A literature review was used in this study. The research questions are as follows:

1. What are the tourist behaviors that encourage overtourism within the scope of the Instagram effect?
2. What has happened in destinations around the world that are subject to overtourism?
3. Is there anything that destination managers/marketers can do to prevent overtourism with the Instagram effect?

Literature review

Overtourism

Overtourism means that, for local people or visitors, there are too many visitors in an area and an unacceptable deterioration in the quality of life or tourist experience in the area. Among the factors that cause excessive tourism, there are reasons such as overcrowding, privatization of public spaces, increase in cruise tourism, and an increase in the number of seasonal and daily visitors. Due to the increase in tourism in the region, tourism speculation and purchasing power parity are decreasing. As a result, people living in highly popular tourist destinations oppose unlimited tourist growth. As a result, when destinations cannot be managed with sustainable tourism growth, attitudes towards tourists emerge in some of the most preferred destinations, especially Venice and Barcelona (Duyar & Bayram, 2019: 347). Goodwin of the Responsible Travel Partnership defines overtourism as destinations where the quality of life or experience at the destination is irreversibly deteriorated when locals or visitors have too many visitors in the area (Goodwin, 2017). Overtourism has emerged as a new structure to study potential dangers to popular destinations worldwide (Sheivachman, 2016). "*Overtourism describes the situation in which the impact of tourism, at certain times and in certain locations, exceeds physical, ecological, social, economic, psychological, and/or political capacity threshold*" (Peeters *et al.*, 2018: 22).

In Figure 5.1, it is emphasized that each destination has an existing 'tourism market mix, volume and growth'. The share of Airbnb offerings in the accommodation sector has an impact on overtourism. Overtourism causes tourism impacts that are not dependent on tourism intensity, environmental pressure and lack of policies. Complex thresholds are available for each domain. Each of these thresholds is not fixed or equal for certain targets. Even if only one of these thresholds is exceeded, the target may return to the overdrive state. Otherwise, the destination will follow the current growth process. If a destination reaches an overtourism situation, it may result in social and/or cultural conflicts, as well as socialization, between visitors and residents. Political reactions include changes to the tourism market mix, volume and growth, as well as a reduction in the carrying capacity of destinations and the multifaceted effects of overtourism (Peeters *et al.*, 2018: 23).

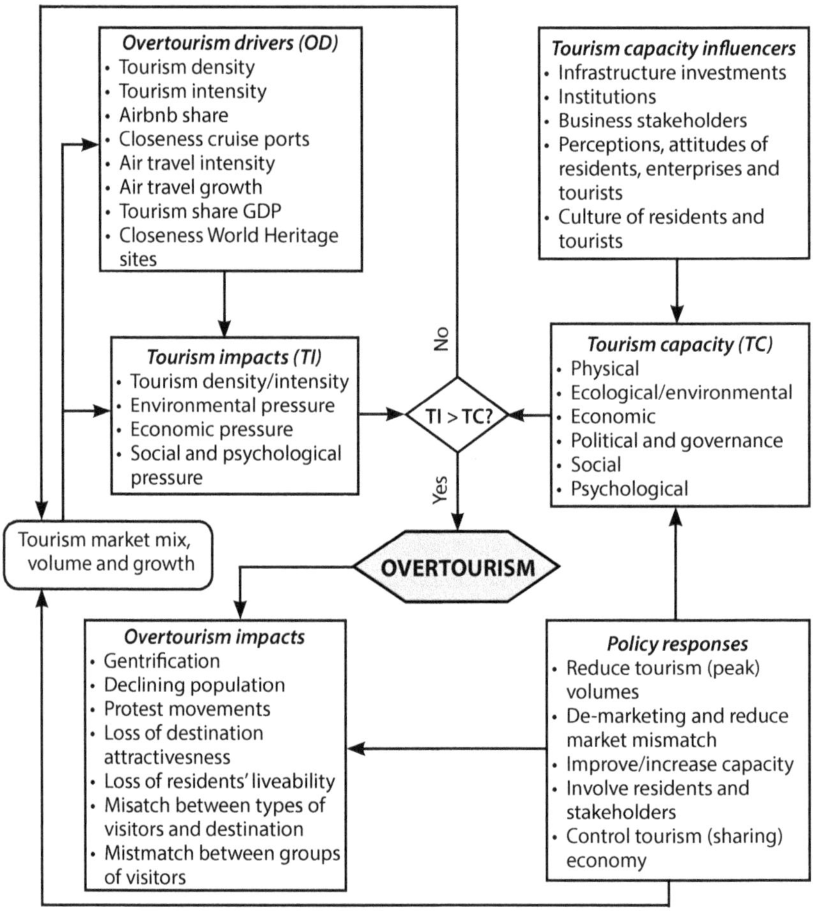

Figure 5.1: Conceptual model of overtourism. Source: (Peeters et al., 2018: 23).

Social media has enabled interaction and communication to grow in a way that has never been seen before in any media technology. The opportunities provided by social media have started to be used as important tools for both consumers and businesses. Especially for tourism businesses and tourists, the use of social media offers significant advantages (Eryılmaz & Yücetürk, 2018: 210). Overtourism is a growing problem. "*Over one billion tourists travel internationally each year, and more than ever, Instagram is determining where and how we travel. People choose destinations based on how good it will look on their Instagram grid, while almost every major city has a roundup of the 'most Instagrammable places to visit'."* (Wayfairer, 2019).

Changes in tourist behavior and destinations exposed to overtourism

Instagram was founded in October 2010 by two entrepreneurs, Kevin Systrom and Mike Krieger. Initially, Instagram, a free photo and video sharing application on social media, allowed its users to use a digital filter on a photo they took and share it with social media services, including Instagram (Onurlubaş & Öztürk, 2018: 989). Headquartered in San Francisco, Instagram is now one of the world's most important social media companies. It was acquired in April 2012 by social media giant Facebook for $1billion (Onurlubaş & Öztürk, 2018: 989). The photo sharing platform Instagram continues to rise. In September 2017, the company recorded 800 million active users, 500 million daily. At the beginning of 2018, the number of users on Snapchat reached 191 million and on Facebook reached 196 billion users, while Instagram reached 1 billion users. (Hürriyet Gazetesi, 2018). In Turkey in June 2019, the audience of social media advertising platforms are based on distribution clicks. Facebook reached 43 million users. Instagram had 38 million users, while Twitter and Snapchat accounted for 10% of social media users (Medya Akademi, 2019).

In social media channels, tourists share their holiday experiences with people they know and do not know, and inform them about the consumption of tourism and create awareness about the product. Thus, the share of social media is important in transforming people who see holiday posts into a potential tourist (Çetinsöz & Akdağ, 2015 as cited in Akyol & Arıca, 2016: 107). In a study conducted by the University of Georgia, the concepts of the 'Instagram effect', 'majority effect (without lag)' and 'dandy effect (contradiction)' were defined for those affected by social media. Some people choose destinations based on current trends, while others try to be cool by choosing non-popular destinations. In both cases, social media decisions affect people's holiday decisions (Karsten, 2021). Various sites, such as the Bucket List (2007) and the Places to See Before You Die, are pushing people to travel to specific destinations to do specific activities. These activities range from things such as diving with fish in the Red Sea, living like a native in Venice, and watching the Northern Lights. The To-Do List concept, associated with a sudden increase in the number of foreign visitors in some of the world's most delicate historical sites, such as Machu Picchu and Venice, commodifies travel experiences and drives people away from different travel pleasures (Bucket List, 2016; Dickinson, 2018 as cited in Duyar & Bayram, 2019: 351).

In the last 10 years, tourism has been driven by various economies on a global scale with the rise of the middle class. This is especially true for emerging economies like China and India. With more disposable income, millions of new travelers start their way to destinations to collect photos for Instagram shares. With low cost airlines like WOW Air and alternative accommodation such as Airbnb, travel also becomes more affordable and accessible, contributing to overtourism (Karsten, 2021). In addition, the emergence of smartphones capable of capturing the highest quality photos and the growth of sharing platforms such as Instagram have made almost every traveler a serious photographer. Certainly, everyone participating in tourism has the right to take his or her own photos, but it also means that tourists are exposed to overtourism by selfie-stick-curious travelers looking for the perfect attraction in places of interest. (Wayfairer, 2019). The hundreds of people holding their palms by the Leaning Tower of Pisa or thousands of people gathering in Angkor Wat to catch the sunrise every morning, are destinations visited to say "Yes, I did it". "*We all take the same travel photos, whether they're digital or print.*" (Wayfairer, 2019).

At a panel discussion on Instagram and Overtourism, organised by We Are Social, Ged Brown stated that as the basis of extreme tourism, air travel is linked to liberalization and the growing middle clas, and that the gap created by the depletion of travel journalism and the lack of objective editorial works are major reasons. In the same session, Besma Whayeb stated that influencers and people traveling only for content play a huge role; while Jen Stubley claimed that we live in an 'Instagramable culture', where people are fascinated by the pictures in the landscape, and new generations are now rushing to Instagramable destinations, competing with each other to capture the most amazing frame. Thus, accelerating extreme tourism is also known in some places as the Instagram/bandwagon/snob effect (Doleman, 2019). Various tourist destinations affected by overtourism are listed below:

- There is evidence that the breaking point has been reached on the Greek island of Santorini due to overtourism. Spectacular sunsets and sea views on the island attract many holidaymakers through Instagram. In 2018, 5.5 million overnight stays were recorded. However, the island is only 76 km^2 and has developed a problem of traffic congestion and overcrowding, with increased water and energy consumption (https://www.tourismdashboard.org).

- Today, the goal management approach to overtourism in Bulgaria still appears to be weak, fragmented, with no clear evidence of a strategic approach to tackling problems. Problems of overtourism continue to manifest, possibly affecting overall environmental degradation and worsening Sunny Beach's target image (Peeters *et al.*, 2018: 221).
- The results of Instagram on social media have reached new levels. *"When a small sunflower farm in Canada experienced overtourism due to Instagram visitors, the police took action to block roads and protect the farm from further destruction."* (Karsten, 2021). When two women began to take a photo shoot at the Love Fountain, Italy's most famous fountain, they fought a violent fight with the negative impact of overtourism (Giuffrida, 2018).
- *"National Parks in the US are not sure how to deal with crowding. Nowadays, overtourism is not clearly evident at Yellowstone national park, but the steady growth in the number of tourists will result in overtourism. ... Currently, the number of tourists is on the rise for visiting national parks and it seems that the management is unable and has no idea on how to deal with over tourism. New transportations systems and infrastructure can indeed, facilitate the increasing number of visitors and accommodate more tourists, however, these are temporary solutions."* (Peeters *et al.*, 2018: 239).
- The number of visitors is expected to increase in Brugge in the near future. Various discount offers and travel programs are encouraging more day tourists. Plans have been made to strengthen the link between the port area (Port of Zeebrugge), expand tourism offers and distribute visitors more evenly and relieve pressure on the Golden Triangle (Papp *et al.*, 2018; Peeters *et al.*, 2018: 160).
- *"There is no question that the 'Instagram effect' on tourism is powerful. For example, the town of Wanaka in New Zealand saw a tourism increase by 14% when promoters invited social media influencers to visit the lakeside city. Now, local residents are somewhat dissatisfied with tax increases proposed to develop tourism infrastructure."* (Tourism Review News, 2019).
- *"Just 800 people visited Norway's picturesque Trolltunga "Troll's Tongue" viewpoint in 2010, while 80,000 people made the hike in 2016. This massive surge of tourism ... was fueled in a large part by social media."* (Karsten, 2021).

- *"Over-tourism and inadequate sewage treatment led to a six-month shutdown of Boracay island in the Philippines beginning last year on the order from president Rodrigo Duterte."* (Tourism Review News, 2019).
- Recent protests against overtourism in Venice and Barcelona illustrate the plight of European cities. Rents have become too expensive for local residents as investors buy apartments to rent to tourists. (Karsten, 2021).
- *"In Iceland. officials are asking visitors to 'please skip' some of the most Instagram-popular and overcrowded spots like Reykjavik's blue lagoon. 2 million tourists arrived in Iceland last year, in contrast to the country's population of less than 350,000. Much of the increase is attributed to social media."* (Tourism Review News, 2019).
- *"Masses looking for the perfect selfie spot are threatening the ancient ruins of Machu Picchu. 1.2 million people visited the site in 2014. This year Peruvian officials instituted a strict ticketing policy, giving visitors access for a limited number of hours."* (Tourism Review News, 2019).
- Thanks to other travelers visiting Bali's Pura Lempuyang and their dream-like attraction on Instagram, this former sacred Balinese Hindu temple has become more popular on social media. *"Known as the 'gateway to heaven', these pictures depict magical water reflections and temple spires in the clouds. In reality, visitors arrive to a massive line of tourists all waiting to get the same picture. A local photographer instructs you to stand at the temple, then puts a mirror beneath your camera to create an illusion of a water reflection."* (Wayfairer, 2019).
- Due to overtourism in remote areas such as the Phi Phi Islands in Thailand, the situation is not bright. *"Because of attention received on social media, the beach received 5,000 visitors per day, causing severe environmental damage. The destination is now closed indefinitely."* (Tourism Review News, 2019).
- *"Recently, I was scrolling through my Instagram feed when I came across a picture tagged as Millaa Millaa Falls, in Australia. The photographer had edited the image to have dazzling blue water. I've been to these waterfalls and they looked nothing like this pretty creation."* (Wayfairer, 2019).
- Peeters et al., 2018, has over 40 case studies of other tourist destinations affected by overtourism..

Due to social media, many tourism destinations and experiences can be in high demand. One of these popular experiences is the Eastern Express, which has become popular in Turkey in recent years. Run by the Turkish Transport Authority, the T.C.D.D, this is a 24-hour train ride experience on the 813 miles route between Ankara and Kars. During the trip, 30,000 content was shared by tourists with the #karsrail tag and more than 18,000 content with the #doguekspresi tag on various social media sites, particularly Instagram. The spread of these posts among more and more users attracted the attention of potential tourists, which led to the intense demand for train tickets. Therefore, the T.C.D.D. Eastern Express is trying to meet the demands by adding additional wagons and offering open tickets for sale one month in advance. In the study conducted by Eryılmaz and Yücetürk (2018), the effect of social media on the reasons why tourists prefer the Eastern Express was investigated by using the interview technique with 20 tourists experiencing the East Express. The results found that the photos and videos shared on Instagram had an important role in the preference of people to the East Express train services (Eryılmaz & Yücetürk, 2018: 210).

Conclusion

Today, more than 1 billion tourists travel internationally and cause overtourism. Social networking is one of the factors that cause overtourism; people's desire to share flawless Instagram photos in tourist destinations (Instagram/bandwagon/snob effect) – how much will be appreciated in Instagram networks, "my life is better than yours", "where you want" – such narcissistic behavior. In addition, the desire to make a list of things to see and do before dying, diving with the fish in the Red Sea, watching the northern lights, spending a few nights in the deserts, such as with bedouins, increases the experience of overtourism. Other factors include, the proliferation of low-cost airlines, the increase in alternative accommodation options such as Airbnb, the emergence of smartphones that shoot perfectly for photographers, and increased Instagramable culture. Such tourist behavior, accommodation, flights and technological changes attract tourists to Instagramable destinations. These developments and changing human behavior have made overtourism inevitable. In this research, various examples are given in the context of world destinations about how today's tourist behaviors turn into overtourism with the Instagram Effect. For destination managers and marketers experiencing difficulties with overtourism, and for responsible travellers, the following recommendations have been developed:

- In all countries, tourist taxes could be increased within the scope of sustainable tourism projects of governments and restrictions on the number of visitors to the region could be set.
- Travelers should refrain from traveling for popular photo and video sharing, and should not post photos of over-tourism-oriented images on Instagram. In particular, travelers should not pave the way for people to turn into potential tourists in destinations that are exposed to over-tourism, by using editing software on photos, to eliminate crowded groups of people and show them as very calm places.
- Traveling photographers should prefer environmentally friendly and sustainable accommodation, use recyclable products and encourage their use on their pages. In addition, it is possible to travel more consciously by visiting destinations that are not very well known in the off-season. While sharing on Instagram, travelers should show an awareness of responsible tourism by not specifying the locations of the destinations, beaches, plateaus or mountains they visit.
- Informative meetings can be organized through zoom or face-to-face at universities by expert academics on issues such as creating a sustainable environment and the harms of over-tourism in order to instill a responsible travel awareness in all travelers.
- Within the framework of cooperation between destination managers and academicians, applications can be developed that prevent over tourism and these applications made available to visitors.
- Destination managers can develop smart city systems in order to prevent congestion by taking advantage of artificial intelligence.

This research has been carried out in detail with a case study of tourist destinations exposed to overtourism by the influence of Instagram and the behaviors of tourists leading to overtourism. Future studies can address the touristic behavior towards the Instagram effect and the problems within the scope of overtourism with quantitative research methods.

Discussion questions

1. How does the Instagram effect change tourist behavior by triggering overtourism?
2. What are the negative outcomes of overtourism for tourism destinations in the world?
3. What can destination marketers do to prevent overtourism caused by Instagram effect?

References

Akyol, M. & Arıca, F.A. (2016) Use of social media in a tourism destination: Users generated content in #Turkeyholiday, *Journal of Tourism Theory and Research,* **2** (2), 98-111.

Bucket List (2016) *Your Dreams, Made Possible,* https://bucketlist.org/ [Accessed: 25 March 2019].

Çetinsöz, B. & Akdağ, G. (2015) Yerli turistlerin tatil sürecinde sosyal medya kullanımı ve tatil kararlarına etkisi, *1 Eurasia International Tourism Congress,* 649-662.

Dickinson, G. (2018) *The Travel Trend to Avoid Before You Die – How The 'Bucket List' Is Ruining The World,* Telegraph. https://www.telegraph.co.uk/travel/comment/how-the-bucket-list-ruined-travel/ [Accessed: 25 March 2019].

Doleman, A. (2019) Instagram versus the reality of overtourism, WeAreSocial.com, https://wearesocial.com/uk/blog/2019/10/instagram-versus-the-reality-of-overtourism [Accessed: 25 October 2019].

Duyar, M. & Bayram, M. (2019) Overtourism and tourismphobia: Evolution of host and tourism relationship, *International Journal of Geography and Geography Education (IGGE),* **40**, 347-362.

Eryılmaz, B. & Yücetürk, C. (2018) Genç turistlerin doğu exspresi seferleri tercihlerinde Instagram'ın Rolü, *Journal of Tourism and Gastronomy Studies,* **6** (4), 210 228.

Goodwin, H. (2017) The challenge of overtourism, *Responsible Tourism Partnership,* **4**, 1-19.

Giuffrida, A. (2018) Selfish selfie-takers cause Trevi fountain fisticuffs, *The Guardian*, August 10th, www.theguardian.com/world/2018/aug/10/selfish-selfie-takers-spark-trevi-fountain-fisticuffs

Hürriyet Gazetesi (2018) Instagram 1 Milyar Kullanıcıya Ulaştı, http://www.hurriyet.com.tr/teknoloji/instagram-1-milyar-aktif-kullaniciya-ulasti-40874508 [Accessed: 18 October 2019].

Karsten, M. (2021) Liked to death: Is Instagram & social media ruining travel? https://expertvagabond.com/instagram-tourism-impact/

Medya Akademi (2019) *Instagram*, http://www.medyaakademi.org/2014/11/09/instag-ramin-kurulusu-ve-tarihi/ [Accessed: 10 October 2019].

İlban, M.O. & Güleç, E. (2018) Sosyal medyanın turistik deneyim üzerindeki etkisini belirlemede sosyal medya kullanma yoğunluğunun aracı rolü, *Journal of Travel and Hospitality Management*, **15** (1), 57-73.

Onurlubaş, E. & Öztürk, D. (2018) Sosyal medya uygulamalarının y kuşağı satın alma davranışı üzerine etkisi: Instagram örneği, *Uluslararası Toplum Araştırmaları Dergisi-International Journal of Society Researches*, **8** (9), 984-1016.

Papp, B., Postma, A. & Koens, K. (2018) *Visitor pressure and events in an urban setting: Understanding and managing visitor pressure in seven European urban tourism destinations*, paper prepared for CELTH: Breda, The Netherlands.

Peeters, P., Gössling, S., Klijs, J., Milano, C., Novelli, M., Dijkmans, C., Eijgelaar, E., Hartman, S., Heslinga, J., Isaac, R., Mitas, O., Moretti, S., Nawijn, J., Papp, B. & Postma, A. (2018) Research for TRAN Committee – overtourism: Impact and possible policy responses, European Parliament, *Policy Department for Structural and Cohesion Policies*, Brussels. http://www.europarl.europa.eu/RegData/etudes/STUD/2018/629184/IPOL_STU(2018)629184_EN.pdf

Sheivachman, A. (2016) Iceland and the Trials of 21st Century Tourism, Skift, https://skift.com/iceland-tourism/ [Accessed: 21 November 2019].

Tourism Review News (2019) Social media boost the over-tourism crisis, https://www.tourism-review.com/social-media-blamed-for-over-tourism-news11005 [Accessed: 22 November 2019].

Wayfairer (2019) Is Instagram really to blame for overtourism? https://www.wayfairertravel.com/inspiration/instagram-and-overtourism/ [Accessed: 25 October 2019].

6 The role of emoji use in destination decision making

Alaattin Basoda, Seden Dogan, and Cihan Cobanoglu

Introduction

Social media is defined as *"a media system that allows users to build two-way simultaneous communication with the emergence of Web 2.0"* (O'Reilly, 2005). The first social media platforms were blogs which were simple web pages that provide information on specific topics in reverse chronological order. The users read the writings and write comments to interact with the blog owners. Interacting through blogs has spread among Internet users quickly. Then, media platforms that allow users to generate their content have emerged. The content generated by users is being followed, reviewed, and shared by other users. This content includes letters, words, sentences, and visual elements (Firat, 2019). The visual elements that have been developed by Shigetaka Kurita in the late 1990s were named *emoji* and *emoticon* (Dogan & Collins, 2019). Emoticons express facial emotions, and emojis covers every area of life, such as animals, food, weather, etc. The emojis and emoticons are enabled for wordless communication or to strengthen the message.

Many companies have discovered social media's benefits and created accounts on different online platforms such as Facebook, Instagram, Twitter, and many others. They use these platforms to promote their products and services by sharing posts, images, and videos — this new form of promoting is named 'social media marketing' . Companies also started to use emojis and emoticons on their promotional posts to attract their followers. Especially, fast food restaurants lead the usage of them. However, there is not much information about how consumers perceive and react to emojis and emoticons. The literature has limited studies in the hospitality and tourism domain about this topic. Vidal *et al.* (2016) conducted a study on tweets about food that used emojis and emoticons. They analyzed 12,260 tweets and found that emojis were used more often than emoticons. Hsieh and Tseng (2015) stated that emoticons in text messages made messaging more fun and enjoyable, thus positively impacting interpersonal communication. Tossell *et al.* (2012) stated that at least one emoticon was found in 4% of text messages.

In most previous studies, emojis were interspersed independently outside (next right, bottom, in text, etc.) of the image's own space (e.g., Das *et al.*, 2019; Eru and Yakin, 2019; Huang *et al.*, 2020). On the contrary, this chapter draws attention to the fact that emojis are positioned directly on the image, not next to and around the image. Therefore, this study is the first to perform such an application. This chapter examines the role of emoji use on images in travelers' destination decision-making process. Travelers' positive affect (PA) and behavioral intentions (BINs) towards the destination [*to search for information about the destination (SI), to visit the destination (VD), to share the post on their social media accounts (SP), to subscribe to the destination's newsletter (SN), and to go on a virtual tour of the destination (GVT)*] and relationships with each other were considered as the steps of their decision-making process related to the destination. As a result, this chapter mainly explores how emoji presence on the promotional social media images differentiates the effects of travelers' reactions (PA) on each intention. Each of the five behavioral intentions was carried out separately, as each demonstrates a different intention towards the destination.

Literature review

Spitzberg (2006: 630) defined computer-mediated communication (CMC) as "*any human symbolic text-based interaction conducted or facilitated through digitally-based technologies.*" Emoticons and emojis are the digital visual communicative elements, named as graphicons, used in CMC (Herring & Dainas, 2017). Emoticons were created in the form of symbols/icons ":-)" and then into pictograms "😊" to express and represent human emotions in CMC in 1982 (Skovholt *et al.*, 2014), later emoticons transformed into emojis. Emojis were designed for mobile phones in Japan in the late 1990s and portrayed as "e=picture, moji=character" (Novak *et al.*, 2015). Emoticons and emojis are facial and body icons that represent nearly nonverbal personal emotions in text-based online communication in digital technology (Tang & Hew, 2019). Evans (2015: 1) defined an emoji as "*an iconic, visual representation of an idea, entity, feeling, status or event, that is used alongside or instead of words in digital messaging and social media.*" Thousands of emojis are used, such as facial expressions, concepts, objects (food, weather, activity, animal, plant, etc.) (Mathews & Lee, 2018).

Emoji is described as a form of visual language (Boothe & Wickstrom, 2017), emotional communication, visual expression symbol, visual paralanguage, visual rhetoric (Bai *et al.*, 2019), and non-linguistic expression (Das, 2020) used in CMC. Emoji is one of the most commonly used online visual elements in digital networks such as social media, instant text messages, and e-mail (Stark & Crawford, 2015). Therefore, emoji usage has become widespread in almost every phase of personal and professional human life by utilizing almost every online communication channel (Kaye *et al.*, 2021). Emojis have become an indispensable part of social media (Wijeratne *et al.*, 2017; Yu & Zhang, 2020). Individuals use emojis rather than text on social media (Abeysinghe *et al.*, 2018), because emojis are the easiest, brief, and fastest way to express one's emotions and thoughts on social media (Yakin & Eru, 2017). While emojis are helpful tools used to understand and improve the social-psychological characteristics of human communication (Gesselman *et al.*, 2019), social media has a direct connection with social psychology (Kende *et al.*, 2015). It can be said that the most effective area for emojis is social media.

On the other hand, Bai *et al.* (2019) stated that emojis have two main functions: emotional (expressing emotions) and semantic (making message meaning). It is an accepted fact that emoticons and emojis are online forms of expressing human emotions visually. Thereby, emotions are at their core (De Angeli *et al.*, 2020; Tang & Hew, 2019). Emojis are tools to analyze users' feelings and perspectives on any object, such as a product (Scherr *et al.*, 2019), and emoticons allow users to express, highlight, and clarify their sentiments in texts (Hogenboom *et al.*, 2013). Many analysts support these views and emphasize that messages or visuals containing emojis are more effective (Novak *et al.*, 2015) and grab the viewer's more emotional attention than those without emojis (Beattie, 2017; Daniel & Camp, 2020). These points illustrate that emojis have various special functions such as awakening emotions (Das *et al.*, 2019), increasing positiveness, empowering self-expression (Eru & Yakin, 2019), creating-reinforcing meanings, evoking-conveying positive feelings (Bai *et al.*, 2019), leaving positive impressions (Wibowo *et al.*, 2017), creating intimacy, providing psychological courtship (Gesselman *et al.*, 2019), lighting the mood, establishing emotional tone (Kaye *et al.*, 2016), displaying personality and differences (Marengo *et al.*, 2017), capturing emotional states (De Angeli *et al.*, 2020), activating a positive effect (Das *et al.*, 2019) and forming interpersonal bonds (Das, 2020) in the viewers in digital social interaction networks. These functions show that emojis differentiate, deepen, and reinforce the meaning of online images or texts, thus leaving a meaningful emotional impact on the viewers during digital communication (Bai *et al.*, 2019). Li *et al.* (2018) also suggest that sensory cues containing emojis used in online reviews about restaurants may make them more enjoyable for readers.

The decision-making process

Due to the nature of tourism, it is known that taxonomies on customer decision-making processes are manifold (Terttunen, 2017). As stated in research conducted by Smallman and Moore (2010), many models related to tourists' decision-making process have been developed. Travel stimuli, personal and social determinants, exogenous variables, motivation, information search, evaluation of alternatives, and decisions have been studied by Schmol, 1977. Awareness, desire, and destination image have been developed by Mathieson and Wall, 1982. Motivation, images, expectation, perception, and preference have been studied by Goodall, 1991. The needs, requests, goals, perceptions, and attitudes have been identified by Middleton and Clarke, 2001. One of the most used models for the tourists' decision-making process in tourism literature is the model of goal-directed behavior (MGB) developed by Perugini and Bagozzi (2001). As indicated at the MGB, the individual's actual behavior is determined by direct perceived behavioral control, frequency of past behavior, and behavioral intention, and indirectly by attitudes, anticipated emotions, subjective norms, perceived behavioral control, and desire (Perugini & Bagozzi, 2001). It is stated that tourists' behavioral intentions are the best cues for understanding the decision-making processes for their actual future behaviors (Lam & Hsu, 2004; Terrah *et al.*, 2020). Therefore, it is understood that the MGB is highly effective in determining tourists' decision-making about destination choice and information searches for traveling to these destinations (Song *et al.*, 2014).

MGB also posits that emotions have an undeniable impact on the tourists' decision-making process, such as evaluating, forming, intention, and choosing (Leone *et al.*, 2005). Supporting this idea, it can be emphasized that emotion is one of the most important antecedents of tourists' behavioral intentions, such as recommendation and revisit (Hosany *et al.*, 2015) because tourists tend to make an emotional connection to the destination (Kim & Stepchenkova, 2017). Empirically, Bagozzi *et al.* (2016) proposed that consumers in the purchasing process consider positive and negative anticipated emotions and found that these emotions affect their purchase intentions. Accordingly, Huseynov *et al.* (2020) also found that tourists' positive anticipated emotions positively affect tourists' desire and intention to visit the destination. Godovykh and Tasci (2020) asserted that the most commonly used component in many studies on the tourist experience is the affective component, affecting feelings, emotions, and mood. Hosany and Gilbert (2010) found that tourists' emotional experiences influence satisfaction from the destination and recommend the destination. Pestana *et al.* (2020) drew attention to psychological factors' influence on the destination choice decision-making process from a different perspective. They found that emotions affect customers' satisfaction with stay and behavioral intentions (revisit and recommend the destination). They emphasized that emotions have a critical impact on the decision-making and purchasing behaviors of tourists. In addition, they suggested that using emotional associations in destination promotion may be a differentiation strategy. They also stressed that visitors' emotional bonds with the destination might strengthen their decision-making process because tourist destinations evoke visitors' emotional responses. Correia, Oliveira, and Butler (2008) also state that emotion is essential in the destination image. They asserted that visitors tend to make emotional evaluations about destinations. They described the consumers' emotional evaluation towards a destination as their emotional response or attachment to it. They also created awareness that emotional image affects the choice of destination.

Moreover, Estrada *et al.* (1994: 286) defined positive affect "*as a pleasant feeling state or good mood.*" Morris *et al.* (2002) stated that emotional response is a powerful determinant of brand attitude, interest, and purchase intention. It is a tool that examines complex feelings involving consumers' reactions to advertisements. Many scholars (Hosany *et al.*, 2020; Loureiro, 2020) asserted that positive effect plays a vital role in travelers' decision-making processes. Based on Loureiro (2020), in this study, PA refers to the image-evoked PA and the degree to which travelers feel cheerful, pleasured, joyful, enthusiastic, delightful, enjoyable, happiness, entertained, comfortable, amazed, astonished, fascinated, inspired, and surprised when they are exposed to images promoting the destination.

With the development of digital technologies, social media has come to the fore and has significantly affected customer behaviors such as online travel information search (Di Pietro *et al.*, 2012). This situation has led to the emergence of critical information-based developments and strategies (searching, planning, sharing, networking, etc.) in the online tourism field. Social media has a critical role in this area (Xiang & Gretzel, 2010). Researchers underlined that social media plays a vital role in consumers' whole travel planning process (before, during, and after, searching, decision making, choice of destination/accommodation) (Fotis *et al.*, 2012). Tham *et al.* (2013) pointed out that social media has a significant cognitive and emotional impact on destination decision making, image, and choice. Zeng and Gerritsen (2014) also stated that social media is a valuable tool to develop strategies for promotion and marketing activities in tourism. Over time, effective social media tools have brought tourism promotion to a higher level. Visuals posted on these are among the most effective methods to promote destinations, reinforce positive destination-oriented perceptions and encourage tourists to choose the destination (Fatanti & Suyadnya, 2015). It is also stated that online photos shared on social media affect customer behaviors on destination choice and the purchasing decision-making process (Terttunen, 2017). Based on the underlined results of many studies, it is clear that social media has effective, supportive and stimulating roles in tourist behaviors such as pre-information seeking, sharing (Höpken *et al.*, 2020; Zeng & Gerritsen, 2014), the way of information seeking, making decision and booking, collecting information, evaluating alternatives, choosing (Gupta, 2019), and intention to visit the destination (Koo *et al.*, 2016).

Previous studies on emojis or emoticons

Emojis or emoticons have been one of the current topics researched in many fields such as linguistics, communication, medicine, behavioral science, psychology, education, marketing, and computer (Bai *et al.*, 2019). Mathews and Lee (2018) explored how many industries (e.g., food and beverage and other services) use emojis as a tool in their marketing strategies to increase consumer engagement. However, studies on the use of emoji in the hospitality and tourism context are rare. Some of these studies reveal that emojis or emoticons influence online reviewer helpfulness (Huang *et al.*, 2020); destination image, brand promotion, visual perception (Shao *et al.*, 2020); visitor's emotional experience (De Angeli *et al.*, 2020); positive Word of Mouth (WOM); intention to purchase and service quality (Ornek, 2019); consumer perception (Xavier, 2018); and e-WOMs (Harris &

Prideaux, 2017). Emojis have been considered the newest promotional tool in digital marketing (Yakin & Eru, 2017) to create awareness and attract the target markets' attraction (Leung & Chan, 2017). As stated in Bezirgan & Kaya (2019), emojis have been widely used as a cross-border communication tool because of having a universal language.

Hence, using emoji can make tourism advertising in social media more noticeable and alluring. Shao *et al.* (2020) found that emojis' design quality as a promotion element is a significant and positive influential factor for both cognitive and affective destination image. They also asserted that using emojis plays an essential role in promoting a destination and effectively encourages people to visit a destination. De Angeli *et al.* (2020) conducted a study to evaluate the emotional experiences (happy or unhappy) of museum visitors and found that emojis can capture emotional states that visitors felt beyond happy dilemmas or not. Ornek (2019) found that mutually using emojis in online customer comments and companies' responses to these comments increased positive WOM, intention to purchase, and expected service quality. Additionally, Xavier (2018) found that using emojis affects customer perceptions about the online message service of the restaurant to book a table.

Jang *et al.* (2009) found that PA affected travelers' motivation and travel intention positively and mediated the effect of novelty-seeking motivation on travel intention. In another study, it was found that ad-evoked PA affected intention to visit the destination (Hosany *et al.*, 2020). Loureiro (2020) also found that PA mediated the effects of destination authenticity on revisit intentions. Therefore, it is seen that many studies were conducted on travelers to examine the impacts of positive affect on their behavioral intentions. However, there is no study examining the effect of emoji-evoked PA on behavioral intentions towards destination.

Studies in other contexts, except tourism, are also striking. For example, Vidal *et al.* (2020) stated that they allowed consumers to add emoji to open-ended product definitions, and as a result, customers' engagement in the product evaluation process increased. They also revealed that consumers attribute a hedonic meaning to products during this process by mainly using positive emojis, hence broadening the range of product-related information. Das *et al.* (2019) examined how emoji use in advertisements influences consumers' PA and purchase intentions and found that emoji presence affects customers' purchasing intentions mediated by PA. Eru and Yakin (2019) found a significant difference between the young people's attitudes towards the emotional advertising message containing emojis and the emotional advertising message without emojis, in social media advertising campaigns. They also stated that emoji marketing had become an essential digital marketing strategy across the world. Daniel and Camp (2020) investigated the relationship between emoji usage and perception of social communication. They found that messages containing original emojis are more understandable and believable than those without emojis or with emojis inappropriate for the context of the messages. Paiva (2018) found that emoji use in consumer-brand e-commerce transactions affects consumers' perception of brand and quality of service. Beattie (2017) found that sources of text messages are rated as more computer-mediated communication competent when they employ emoji, than when using verbal-only messages. Hill (2016) found that positive valence-based messages containing emoticons affect purchase intention positively.

Methodology

To achieve the chapter's main aim, two questions were produced; (1)"*How does the travelers' PA evoked by the images with or without emoji towards the destination influence their BINs?*", and (2)"*Does the emoji presence moderate and differentiate (versus absence) the effects of PA on BINs significantly?*. Therefore, an online scenario-based quick questionnaire was created with four promotional images about the destination (Amasra/Turkey) that can be posted on social media by a tour operator or a destination marketing organization to attract potential tourists: one with pure image, one with pure image and emoticons in the description text, one with emojis on the image, and one with emojis on the image and emoticons in the descriptive.

Figure 6.1: (a) Emoji absence on the image-1. (b) Emoji absence on the image-2.

Figure 6.1: (c) Emoji presence on the image-1. (d) Emoji presence on the image-2

The participants were randomly exposed to one of the four images and then answered a self-administered questionnaire by scrutinizing the image. In the first part of the questionnaire, they were asked five questions related to behavioral intentions, which were aimed to determine how inclined they were to search for information about the destination (SI), to visit the destination (VD), to share the post on their social media accounts (SP), to subscribe to the destination's newsletter (SN), and to go on a virtual tour of the destination (GVT). All items were rated from 1 (not at all) to 5 (very much). The second part of the questionnaire included the question, "*While exposed to the post, how did you feel?*" which was asked to reveal how the viewers felt when they saw the images. They were given a ready set of PAs and asked them to rate each affected item from 1 (not at all) to 7 (very much). A ready PA-set has consisted of "*cheerful, pleasured, joyful, enthusiastic, delightful, enjoyable, happiness, entertained, comfortable, amazed, astonished, fascinated, inspired and surprised*" (Kumar & Nayak, 2014). The last part of the survey consisted of demographical questions.

The general design of the study was inspired by Das *et al.* (2019). The questionnaire was sent to three academics (one full professor and two assistant professors) who were well-known in research methods. Based on their recommendations, the questionnaire was finalized. The primary data was collected through a self-administrated online survey developed using Qualtrics. The survey was distributed via Amazon Mechanical Turk (mTurk). Scholars (Buhrmester *et al.*, 2011; Goodman *et al.*, 2013; Stritch *et al.*, 2017) have mentioned that mTurk is a reliable source to collect data. In total, 1301 returns was gathered from mTurk. After the deletion of outliers and missing data, 1102 (male=561, female=541) respondents remained for analysis. Since the primary purpose is to determine the role of emoji presence on the images in the decision-making process, for use in analysis, the four scenario-based images applied in the questionnaire were grouped under two groups named as the presence (n=551; male=295, female=256) and absence (n=551, male=266, female=285) of emoji on the images. To analyze the relationships between PA and BINs (SI, VD, SP, SN ns GVT), linear regression analyses were performed. Then the moderation model 1 was utilized by the PROCESS software with 5000 bootstrap samples and a 95% confidence interval (Hayes, 2013) to explore how the role of emoji presence (versus absence) in the relationships between PA and BINs (SI, VD, SP, SN, and GVT). The PA scale's reliability was evaluated and found to be 97α for total, 98α for emoji presence, and 96α for emoji absence (Peterson, 1994).

Results

For the first question, linear regression analyses were performed for each group. As seen in Table 6.1, results show that PA has a significant and positive impact on each behavioral intention in both presence and absence groups. The results conspicuously indicate that PA has a high-scored explanation of the change in all behavioral intentions for the emoji presence group. In that case, as it is clearly understood from the high-value regression coefficients of the emoji presence group, it can be said that the posts containing emojis on the images strengthen and enhance the effect of PA on BINs. However, statistical analysis is needed to determine if these differences are significant. It will be possible to see these situations under the subsequent analysis.

Coefficients	SI		VD		SP		SN			
	EP	EA	EP	EA	EP	EA	EP	EA	EP	EA
R	0,67	0,47	0,64	0,50	0,58	0,51	0,53	0,45	0,62	0,44
R^2	**0,45**	0,22	**0,41**	0,25	**0,33**	0,26	**0,28**	0,20	**0,38**	0,19
ΔR^2	0,45	0,22	0,40	0,25	0,33	0,26	0,28	0,20	0,38	0,19
F	448,92*	153,79*	375,75*	182,78*	271,81*	193,11*	216,24*	139,40*	336,47*	133,07*
B	**0,60**	0,48	**0,59**	0,58	**0,67**	0,79	**0,61**	0,69	**0,60**	0,56
β	**0,67**	0,47	**0,64**	0,50	**0,58**	0,51	**0,53**	0,45	**0,62**	0,44
t	21,19*	12,40*	19,38*	13,52*	16,49*	13,90*	14,71*	11,81*	18,34*	11,53*

Table 6.1: The effect of PA on BINs (in two groups)

*p<0.001, Groups = **EP**: *Emoji Presence,* **EA**: *Emoji Absence*; Behavioral intentions = **SI**: *search for information about the destination,* **VD**: *visit the destination,* **SP**: *share the post on their social media accounts,* **SN**: *subscribe to the destination's newsletter,* **GVT**: *go on a virtual tour of the destination.*

For the second question, it was found that the overall moderating model was significant only for the SI [$F(3,1098)=205,11$, $p=.00$, $R^2=.36$, $R=.60$, $MSE=.60$]. As per expected, emoji presence moderated the impact of PA on SI, and the interaction emoji presence with PA in relation to SI is significant ($R^2_{change}=.00$, $F(3,1098)=5,98$, $p=.01$). According to the results, the significant predictors of SI were PA ($b_1=.48$, $se=.04$, $t(1098)=12,50$, $p=.00$, $LLCI=.41$, ULCI:.56), emoji presence ($b_2=-.73$, $se=.28$, $t(1098)=-2,60$, $p=.01$, $LLCI=-1.28$, ULCI=-.18) and interaction (emoji presence*affect) ($b_3=.12$, $se=.05$, $t(1098)=2,44$, $p=.01$, $LLCI=.02$, ULCI=.21). Additionally, the results indicated that the effect of PA on SI was also much stronger in emoji presence ($b=.60$, $se=.03$, $t(1098)=21,01$, $p=.00$, $LLCI=.54$, ULCI=.66) than absence ($b=.48$, $se=.03$, $t(1098)=12,50$, $p=.00$, $LLCI=.41$, ULCI=.56). However, the interaction of emoji presence and PA in relation to VD ($R^2_{change}=.00$, $F(3,1098)=.06$, $p=.80$), SP ($R^2_{change}=.00$, $F(3,1098)=3,03$, $p=.08$), SN ($R^2_{change}=.00$, $F(3,1098)=1,20$, $p=.27$), and GVT ($R^2_{change}=.00$, $F(3,1098)=.71$, $p=.40$) were not significant.

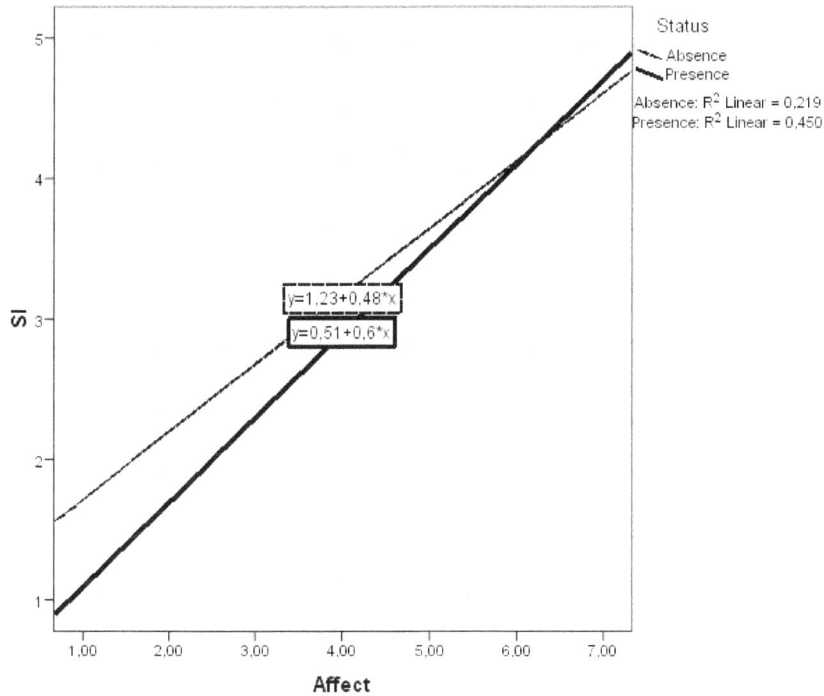

Figure 6.2: Simple slope graph for the moderating role of emoji presence

Conclusion

The literature review showed that emoji/emoticon-related studies in the hospitality and tourism industry are very limited. In this context, this study may significantly contribute to the existing hospitality and tourism literature. In this study, the emoji-evoked positive effect on the travelers contributed to their intention to search for further information about the destination. This result supports some findings in tourism and non-tourism literature. For example, using emojis/emoticons on advertisements and commercials significantly affects customers' purchase behavior (Das *et al.*, 2019; Shao *et al.*, 2020). The positive moderator effect in the relationships between positive affect and information search for the destination was also higher for emoji presence than absence. Hence, this chapter suggests that the characteristics of positive effect evoked by emoji and evoked by pure image may have a different influence on every step of tourists' decision-making process regarding destination via social media.

As a focus inference, it can be said that the emoji-evoked positive effect activates the cognitive process by arousing curiosity about the destination in travelers, which in turn increases the intention to search for information about the destination. Therefore, the emoji is a driving and stimulating force that encourages travelers to seek information about their destination rather than their other intentions. From another aspect, what is commonly known is that emojis emphasize the information in the text. However, this study revealed that emojis strengthen travelers' tendency to seek information about the destination by enriching the visual information about the destination on the images. Consequently, emojis have some remarkable features that enrich visual information and lead viewers to search for more. As per this study, hospitality and tourism organizations may build a successful and interactive communication with travelers to attract them by using emoji/emoticons on their promotional materials, especially on the images.

This study has several limitations like many others. One is that it was used only positive emojis on the images and emoticons in the text. In the survey design, the pure or emoji-contained text was given next to the images. In this way, the effectiveness of emojis seems to decrease. This study suggests that, by using emojis/emoticons in the text on the promotional images (not next to images, but on), more comprehensive scientific studies need to be done for more transparent results on this subject.

Further studies may use negative or neutral emojis/emoticons in the text that is on the image. Another limitation is that we used two emojis on the images. Further studies may use more or fewer emojis. As a part of the survey design, the ready PA set was given to the participants, and they scored with the appropriate rating. Future studies may make PA open-ended so that participants can identify the effects themselves. Another limitation is that the text consisted only of the destination name as Amasra/Turkey. In future studies, texts defining the destination may be included.

On the other hand, we created four different scenarios to show the participants. More complicated scenarios to understand the phenomena can be created. The most critical issue is to find the best emoji(s)/emoticon(s) to attract travelers. Further studies may also focus on different variables to determine the strengths and weaknesses of emojis/emoticons in online communication between customers and companies. Research may be done with more advanced methods such as Brain Data Measuring Hardware and Software to get more transparent results. Also, as Eru and Yakin (2019) stated, emojis are more effective than traditional methods when used simultaneously and with the right target audience. That means that emoji uses, time, and the target audience should be chosen carefully. Lastly, further studies may examine the relationships between emoji presence/absence on the image and the travelers' goal-directed behaviors towards the destination by developing the extended model of goal-directed behavior.

Discussion questions

1. Which emojis or emoticons used on a promotional post can create a positive attitude?
2. Which emojis or emoticons used on a promotional post can create a negative attitude?
3. Can cultural emojis or emoticons on a promotional post be more impactful than the regular ones?
4. How many emojis or emoticons should be used on a promotional post?

Acknowledgment: The authors acknowledge the support from the University of South Florida M3 Center for Hospitality Technology and Innovation.

References

Abeysinghe, S., Manchanayake, I., Samarajeewa, C., Rathnayaka, P., Walpola, M.J., Nawaratne, R., Bandaragoda, T. & Alahakoon, D. (2018) Enhancing decision making capacity in tourism domain using social media analytics, *paper at the Conference on Advances in ICT for Emerging Regions (ICTer)*, 27-28 September, Colombo, Sri Lanka.

Bagozzi, R., Belanche, D., Casaló, L.V. & Flavián, C. (2016) The role of anticipated emotions in purchase intention, *Psychology & Marketing*, **33** (8), 629-645.

Bai, Q., Dan, Q., Mu, Z. & Yang, M. (2019) A systematic review of emoji: current research and future perspectives, *Frontiers in Psychology*, **10**.

Beattie, A.J. (2017) *Interpersonal impressions of emoji use in computer-mediated decision making*, unpublished master thesis, Scholar Works of Western Michigan University.

Bezirgan, M. & Kaya, İ. (2019) Digital communications and emoji use in tourism enterprises, *Theory and Practice in Social Sciences*, **10**, 556-569.

Boothe, D. & Wickstrom, C. (2017) ESOL learners must confront diverging language pathways between social media and English for specific purposes, *10th International Conference of Education, Research and Innovation*, 16-18 November, Seville, Spain.

Buhrmester, M., Kwang, T. & Gosling, S.D. (2011) Amazon's Mechanical Turk: A new source of inexpensive, yet high-quality, data?, *Perspectives on Psychological Science*, **6** (1), 3–5.

Correia, A., Oliveira, N. & Butler, R. (2008) First-time and repeat visitors to Cape Verde: The overall image, *Tourism Economics*, **14**, 185–203.

Daniel, T.A. & Camp, A.L. (2020) Emojis affect processing fluency on social media, *Psychology of Popular Media*, **9** (2), 208–213.

Das, A. (2020) How has the coronavirus (COVID-19) pandemic affected global emoji usage?, *Journal of Human Behavior in the Social Environment*, **31** (1-4), 425-434.

Das, G., Wiener, H.J.D. & Kareklas, I. (2019) To emoji or not to emoji? Examining the influence of emoji on consumer reactions to advertising, *Journal of Business Research*, **96**, 147-156.

De Angeli, D., Kelly, R. M. & O'Neill, E. (2020) Beyond happy-or-not: Using emoji to capture visitors' emotional experience, *Curator: The Museum Journal*, **63** (2), 167-191.

Di Pietro, L., Di Virgilio, F. & Pantano, E. (2012) Social network for the choice of tourist destination: attitude and behavioural intention, *Journal of Hospitality and Tourism Technology*, **3** (1), 60-76.

Dogan, S. & Collins, A. (2019) Does communication with emoticons/emojis work efficiently? Evidence from Turkey, *Advances in Global Business and Economics*, **2**, 396-403.

Eru, O. & Yakın, V. (2019) Research on determining perceptions and attitudes towards emoji use in digital marketing campaigns, *Çukurova Universitesi Sosyal Bilimler Enstitusu Dergisi*, **28** (2), 83-100.

Estrada, C.A., Isen, A.M. & Young, M.J. (1994) Positive affect improves creative problem solving and influences reported source of practice satisfaction in physicians, *Motivation and Emotion*, **18**, 285–299.

Evans, V. (2015) No, the rise of the emoji doesn't spell the end of language, *The Conversation*, Available at: http://theconversation.com/no-the-rise-of-the-emoji-doesnt-spell-theend-of-language-42208 [Accessed: n.d.].

Fatanti, M.N. & Suyadnya, I.W. (2015) Beyond user gaze: How Instagram creates tourism destination brand? *Procedia - Social and Behavioral Sciences*, **211**, 1089–1095.

Firat, D. (2019) YouTube advertising value and its effects on purchase intention, *Journal of Global Business Insights*, **4** (2), 141-155.

Fotis, J., Buhalis, D. & Rossides, N. (2012) Social media use and impact during the holiday travel planning process, In M. Fuchs, F. Ricci, and L. Cantoni (eds.), *Information and Communication Technologies in Tourism 2012*, Vienna: Springer, pp. 13-24.

Gesselman, A.N., Ta, V.P. & Garcia, J.R. (2019) Worth a thousand interpersonal words: emoji as affective signals for relationship-oriented digital communication, *PLoSONE*, **14** (8), 1-14.

Godovykh, M. & Tasci, A.D.A. (2020) Customer experience in tourism: A review of definitions, components, and measurements, *Tourism Management Perspectives*, **35**, 1–19.

Goodall, B. (1991) Understanding holiday choice, in C. Cooper (ed.), *Progress in Tourism, Recreation and Hospitality Management*, London: Belhaven, pp. 59–77.

Gupta, V. (2019) The influencing role of social media in the consumer's hotel decision-making process, *Worldwide Hospitality and Tourism Themes*, **11** (4), 378–391.

Goodman, J.K., Cryder, C.E. & Cheema, A. (2013) Data collection in a flat world: The strengths and weaknesses of Mechanical Turk samples, *Journal of Behavioral Decision Making*, **26** (3), 213–224.

Harris, A. & Prideaux, B. (2017) The potential for eWOM to affect consumer behaviour in tourism, in S.K. Dixit (ed.), *The Routledge Handbook of Consumer Behaviour in Hospitality and Tourism*, Abingdon: Routledge, pp. 366-376.

Hayes, A.F. (2013) *Introduction to Mediation, Moderation, and Conditional Process Analysis: A regression-based approach*. New York: Guilford Press.

Herring, S. & Dainas, A. (2017) 'Nice picture comment!' Graphicons in Facebook comment threads, *paper at the 50th Hawaii International Conference on System Sciences*, 4-7 January, Hawaii, USA.

Hill, J.H. (2016) *The impact of emojis and emoticons on online consumer reviews, perceived company response quality, brand relationship, and purchase intent*, unpublished doctoral thesis, University of South Florida.

Hogenboom, A., Bal, D., Frasincar, F., Bal, M., de Jong, F. & Kaymak, U. (2013), Exploiting emoticons in sentiment analysis, *28th Annual ACM Symposium on Applied Computing*, 18-22 March, Coimbra, Portugal.

Höpken, W., Müller, M., Fuchs, M. & Lexhagen, M. (2020) Flickr data for analysing tourists' spatial behaviour and movement patterns, *Journal of Hospitality and Tourism Technology*, **11** (1), 69-82.

Hosany, S. & Gilbert, D. (2010) Measuring tourists' emotional experiences toward hedonic holiday destinations, *Journal of Travel Research*, **49** (4), 513–526.

Hosany, S., Prayag, G., Deesilatham, S., Caušević, S. & Odeh, K. (2015) Measuring tourists' emotional experiences: Further validation of the destination emotion scale, *Journal of Travel Research*, **54** (4), 482–495.

Hosany, S., Buzova, D. & Sanz-Blas, S. (2020) The influence of place attachment, ad-evoked positive affect, and motivation on intention to visit: Imagination proclivity as a moderator, *Journal of Travel Research*, **59** (3), 477–495.

Hsieh, S.H. & Tseng, T.H. (2015) Playfulness in mobile instant messaging: Examining the influence of emoticons and text messaging on social interaction, *Computers in Human Behavior*, **69**, 405-414.

Huang, G.H., Chang, C.T., Bilgihan, A. & Okumus, F. (2020) Helpful or harmful? A double-edged sword of emoticons in online review helpfulness, *Tourism Management*, **81.**

Huseynov, K., Costa Pinto, D., Maurer Herter, M. & Rita, P. (2020) Rethinking emotions and destination experience: An extended model of goal-directed behavior, *Journal of Hospitality & Tourism Research*, **44** (7), 1153–1177.

Jang, S., Bai, B., Hu, C. & Wu, C.-M. E. (2009) Affect, travel motivation, and travel intention: A senior market, *Journal of Hospitality and Tourism Research*, **33** (1), 51–73.

Kaye, L.K., Rodriguez-Cuadrado, S., Malone, S.A., Wall, H.J., Gaunt, E., Mulvey, A.L. & Graham, C. (2021) How emotional are emoji?: Exploring the effect of emotional valence on the processing of emoji stimuli, *Computers in Human Behavior*, **116.**

Kaye, L.K., Wall, H.J. & Malone, S.A. (2016) Turn that frown upside-down: A contextual account of emoticon usage on different virtual platforms, *Computers Human Behavior*, **60**, 463–467.

Kende, A., Ujhelyi, A., Joinson, A. & Greitemeyer, T. (2015) Putting the social (psychology) into social media, *European Journal of Social Psychology*, **45**, 277–278.

Kim, H. & Stepchenkova, S. (2017) Understanding destination personality through visitors' experience: a cross-cultural perspective, *Journal of Destination Marketing & Management,* **6** (4), 416–25.

Koo, C., Joun, Y., Han, H. & Chung, N. (2016) A structural model for destination travel intention as a media exposure: Belief-Desire-Intention model perspective, *International Journal of Contemporary Hospitality Management,* **28** (7), 1338-1360.

Kumar, V. & Nayak, J.K. (2014) Destination personality: Scale development and validation, *Journal of Hospitality & Tourism Research,* **42** (1), 3-25.

Lam, T. & Hsu, C.H.C. (2004) Theory of planned behavior: Potential travelers from China, *Journal of Hospitality and Tourism Research,* **28** (4), 463–482.

Leone, L., Perugini, M. & Bagozzi, R. (2005) Emotions and decision making: Regulatory focus moderates the influence of anticipated emotions on action evaluations, *Cognition & Emotion,* **19** (8), 1175-1198.

Leung, C.H. & Chan, W.T.Y. (2017) Using emoji effectively in marketing: An empirical study, *Journal of Digital & Social Media Marketing,* **5** (1), 76-95.

Li, H., Wang, C.R., Meng, F. & Zhang, Z. (2018) Making restaurant reviews useful and/or enjoyable? The impacts of temporal, explanatory, and sensory cues, *International Journal of Hospitality Management,* **83**, 257-265.

Loureiro, S.M.C. (2020) How does the experience and destination authenticity influence 'affect'?, *Anatolia: An International Journal of Tourism and Hospitality Research,* **31** (3), 449-465.

Marengo, D., Giannotta, F. & Settanni, M. (2017) Assessing personality using emoji: An exploratory study, *Personality and Individual Differences,* **112**, 74–78.

Mathews, S. & Lee, S. (2018) Use of emoji as a marketing tool: An exploratory content analysis, *Fashion, Industry and Education,* **16** (1), 46–55.

Mathieson, A. & Wall, G. (1982) *Tourism: Economic, Physical and Social Impacts,* New Jersey: Prentice Hall.

Middleton, V.T.C. & Clarke, J.R. (2001) *Marketing in Travel and Tourism* (3rd ed.). London: Butterworth-Heinemann.

Morris J.D., Woo, C., Geason, J.A. & Kim, J. (2002) The power of affect: Predicting intention, *Journal of Advertising Research,* **42** (3), 7-17.

Novak, P.K., Smailović, J., Sluban, B. & Mozetič, I. (2015) Sentiment of emojis, *PloS One,* **10** (12), e0144296.

O'Reilly, T. (2005) *What is Web 2.0,* oreilly.com/pub/a/web2/archive/what-is-web-20.html [Accessed: n.d.].

Ornek, E. (2019) *The impact of emoji use in online consumer reviews and company responses,* unpublished M.B.A. thesis, Istanbul Technical University.

Paiva, N.E.B.M.M. (2018) *Using emoji in an e-commerce context: Effects in brand perception, quality of service and intention to recommend,* unpublished master thesis, IUL School of Social Sciences.

Perugini, M. & Bagozzi, R.P. (2001) The role of desires and anticipated emotions in goal-directed behaviors: A model of goal-directed behavior, *British Journal of Social Psychology,* **40**, 79 – 98.

Pestana, M.H., Parreira, A. & Moutinho, L. (2019) Motivations, emotions and satisfaction: The keys to a tourism destination choice, *Journal of Destination Marketing and Management,* **16**, 1–9.

Peterson, R.A. (1994) A meta-analysis of Cronbach's coefficient alpha, *Journal of Consumer Research,* **21**, 381-391.

Scherr, S.A., Polst, S., Muller, L., Holl, K. & Elberzhager, F. (2019) The perception of emojis for analysing app feedback, *International Journal of Interactive Mobile Technologies,* **13** (2), 19–36.

Schmoll, G. (1977) *Tourism Promotion,* London: Tourism International Press.

Shao, J., Yi, S., Shen, Y., Gretzel, U. & Joppe, M. (2020) Research on the influence of emoji communication on the perception of destination image: The case of Finland, *Travel and Tourism Research Association: Advancing Tourism Research Globally,* https://scholarworks.umass.edu/ttra/2020/research_papers/19/

Skovholt, K., Gronning, A. & Kankaanranta, A. (2014), The communicative functions of emoticons in workplace e-mails:-), *Journal of Computer-Mediated Communication,* **19**, 780-797.

Smallman, C. & Moore, K. (2010) Process studies of tourists' decision-making, *Annals of Tourism Research,* **37** (2), 397–422.

Song, H., You, G.-J., Reisinger, Y., Lee, C.-K. & Lee, S.-K. (2014) Behavioral intention of visitors to an oriental medicine festival: An extended model of goal directed behavior, *Tourism Management*, **42** (1), 101-113.

Spitzberg, B.H. (2006) Preliminary development of a model and measure of computer-mediated communication (CMC) competence, *Journal of Computer Mediated Communication*, **11** (2), 629–666.

Stark, L. & Crawford, K. (2015) The conservatism of emoji: Work, affect, and communication, *Social Media and Society*, **1** (2), 1–11.

Stritch, J.M., Pedersen, M.J. & Taggart, G. (2017) The opportunities and limitations of using Mechanical Turk (MTurk) in public administration and management scholarship, *International Public Management Journal*, **20** (3), 489–511.

Tang, T. & Hew, K.F. (2019) Emoticon, emoji, and sticker use in computer-mediated communication: A review of theories and research findings, *International Journal of Communication*, **13**, 2457–2483.

Terrah, A., Wildes, V. & Mistry, T.G. (2020) Antalya's tourist security: A gap analysis of expectations vs perceptions, *Journal of Global Business Insights*, **5** (2), 150-168.

Terttunen, A. (2017) *The Influence of Instagram on Consumers' Travel Planning and Destination Choice*, unpublished master thesis, Haaga-Helia, University of Applied Science.

Tham, A., Croy, G. & Mair, J. (2013) Social media in destination choice: Distinctive electronic word of mouth dimensions, *Journal of Travel & Tourism Marketing*, **30** (1-2), 144-155.

Tossell, C.C., Kortum, P., Shepard, C. & Barg-Walkow, L.H. (2012) A longitudinal study of emoticon use in text messaging from smartphones, *Computers in Human Behavior*, **28** (2), 659-663.

Vidal, L., Ares, G., Le Blond, M., Jin, D. & Jaeger, S.R. (2020) Emoji in open-ended questions: A novel use in product research with consumers, *Journal of Sensory Studies*, **35** (6), e12610.

Vidal, L., Ares, G. & Jaeger, S.R. (2016) Use of emoticon and emoji in tweets for food-related emotional expression, *Food Quality and Preference*, **49**, 119-128.

Wibowo, M.R.F., Ats-Tsiqoh, R., Sangadah, S., Komala, E.S. & Utomo, A.B. (2017), The effect of emoji on person perception, *UI Proceedings on Social Science and Humanities*, **1** (2017).

Wijeratne, S., Balasuriya, L., Sheth, A.P. & Doran, D. (2017) A semantics-based measure of emoji similarity, *paper at the International Conference on Web Intelligence*, August 23-26, Leipzig, Germany, 646–653.

Xavier, S.C.C. (2018) *With or without emoji?: Impact of the use of emojis on online service booking on consumer perception*, unpublished master thesis, IUL School of Social Sciences.

Xiang, Z. & Gretzel, U. (2010) Role of social media in online travel information search, *Tourism Management*, **31**, 179-188

Yakin, V. & Eru, O. (2017) An application to determine the efficacy of emoji use on social marketing ads, *International Journal of Social Sciences and Education Research*, **3** (1), 230–240.

Yu, C.E. & Zhang, X. (2020 The embedded feelings in local gastronomy: A sentiment analysis of online reviews, *Journal of Hospitality and Tourism Technology*, **11** (3), 461-478.

Zeng, B. & Gerritsen, R. (2014), What do we know about social media in tourism? A review, *Tourism Management Perspectives*, 10, 27–36.

7 Content marketing: Definition, cases, and helpful tactics

Linchi Kwok

Introduction

Content marketing, a means of digital marketing, is a relatively new concept and practice, although marketing has been long rooted in business. It was not until the 21st century that digital marketing emerged from and evolved with internet-based technology and the digital economy (Wymbs, 2011). Today, any marketing methods conducted through electronic devices can be referred to as digital marketing (AMA, n.d.).

While there are many different aspects of digital marketing, including social media marketing, search engine optimization (SEO), search engine marketing (SEM), pay-per-click advertising, affiliate marketing, blogger, e-mail marketing, mobile marketing, etc., content marketing stands out as the highly preferred channel that induces consumers to make a purchasing decision online (Ramesh & Vidhva, 2019). Content marketing plays an increasingly important role in business and receives significant attention in marketing research.

It will be useful at this point to review content marketing to clarify the misconceptions of the subject and highlight some of the significant research findings from relevant literature. A clear understanding of content marketing, a relatively new and still evolving concept, will be helpful for the practitioners who want to advance effective tactics at work. Scholars and students who are interested in content marketing may also be able to spark new ideas as they seek a deeper understanding of the domain. This conceptual review is hence conducted to answer four questions.

1. What is content marketing?
2. Why is it highly relevant to hospitality and tourism businesses?
3. What are the significant impacts of content marketing on hospitality and tourism businesses?
4. What effective content marketing strategies can a hospitality and tourism business take?

Literature review

According to the American Marketing Association, content marketing is "*the publishing and distribution of text, video, or audio materials to customers online. Blogs, videos, and podcasts are common ways for businesses to engage in content marketing.*" (AMA, n.d.) The Content Marketing Institute (n.d.) defines content marketing as "*a strategic marketing approach focused on creating and distributing valuable, relevant, and consistent content to attract and retain a clearly-defined audience – and, ultimately, to drive profitable customer action.*" In a business-to-business context, content marketing "*involves creating, distributing and sharing relevant, compelling and timely content to engage customers at the appropriate point in their buying consideration processes, such that it encourages them to convert to a business building outcome*" (Jävine & Taiminen, 2016).

Regardless of its forms and distribution channels, the core information presented in content marketing is usually brand- or product-related. Some typical content marketing examples include infographics, webpage, podcasts, videos, and books (Steimle, 2014).

It is critical to distinguish 'content marketing' from 'digital content marketing'. These two terms look almost the same. Some research reports and studies might even mistakenly use these two terms interchangeably. Digital content marketing is essentially about the marketing of digital products.

The term 'digital content' is used to substitute 'digital product', 'electronic information products', 'information goods' and 'virtual products', which are bit-based objects that can be distributed through electronic channels (Koiso-Kanttila, 2004). Typical examples of digital content include e-books, e-journal articles, ringtones, and pay-per-view or pay-for-download videos. Digital content marketing is hence defined as "*the management process responsible for identifying, anticipating, and satisfying customer requirements profitably in the context of digital content, or bit-based objects distributed through electronic channels*" (Rowley, 2008: 235).

The hospitality and tourism is a significant component of the service industry. What is offered in the hospitality and tourism industry also carries the unique characteristics of a service product, challenging marketers to address the issues of intangibility, inseparability of production and consumption, heterogeneity of quality, and perishability (De Chernatony & Segal-Horn, 2001). Thanks to content marketing strategies, hospitality and tourism companies can find ways to address some of those challenges, as illustrated in Table 7.1.

Table 7.1: Content marketing and the unique characteristics of a service product

Definition/description	**Possible solutions w/ content marketing strategies**
Intangibility	
Travelers want to enjoy the experience and the process, although tangible goods (e.g., food items) and the facility also play an essential role.	The experience and service process can be vividly described and documented into narratives. Or, they can be captured and presented in a visual format, such as pictures and videos. This kind of content can help hospitality and tourism businesses present authentic experiences to their potential customers.
Inseparability	
Travelers enjoy the experience as the service product is produced and delivered.	Consumers can share their experiences with their networks during the service delivery process (e.g., IGTV & Facebook Live), allowing their followers to participate virtually and be part of the experience. Moreover, such experiences can be shared afterward.
Heterogeneity	
Variations in service quality exist, e.g., differences in consumer perceptions and who offers/delivers the service.	Businesses can influence consumers' perceived quality of a service product by communicating the same content or consistent marketing messages with their target customers. Meanwhile, they can also address and respond to the service failure issues identified from online reviews (one type of user-generated content) to keep consistent service quality.
Perishability	
Travelers must enjoy the experience and process in a specific timeframe.	Potential consumers can better understand other consumers' experiences about a hospitality and tourism product by reading the user- or brand-generated content. Consumers can also retrieve or 'relive'their past experience by reviewing what they documented, captured, and stored themselves.

The significant impacts of content marketing on hospitality and tourism: Some real cases

According to an HVS report, a hotel can increase its price (average daily rate or ADR) by 11.2% while keeping the same occupancy or market share if the hotel can improve one-point of its user generated content (UGC), such as its online review ratings on TripAdvisor.com (Bagley & Lanz, 2014). There are many real cases about how hospitality and tourism companies benefitted from content marketing strategies. Here are a few examples.

Hotel examples

Many luxury hotels, such as The Four Seasons and St. Regis, have a long history of publishing in-room printed magazines, which can also be accessed online. These magazines usually feature brand- or product-related content, such as the exclusive experiences offered in their unique properties, food and beverage offerings, local flavors, things to do in a tourist destination, wellness tips for healthy living and lifestyle, ideas for wedding and family vacation, and more. In this case, content marketing does not limit to the usage of digital media.

In another case, Sanderling Resort saw a year-to-year increase of the property's visual content on Instagram and Pinterest of 400% in 2015, citing their Millennials customers, who loved sharing photos and stories on Instagram (Smith, 2016). The resort also published 115 blogs in 2015 to highlight the weddings, the destination, and things to do in the resort.

Likewise, the corporate marketing team at The Four Seasons Hotels and Resorts reported a great success a few years ago about their content marketing strategies on Pinterest. According to a Pinterest's case study about Four Seasons (Shankman, 2014), the hotel chain shared such lifestyle content as food, fashion, and travel in general, in addition to its property photos or promotions. The chain observed ten times more daily average visitors and 17 times more daily average clicks to the hotel's website. Furthermore, Shankman (2014) illustrated how Pinterest helped The Four Seasons turn visual content into real profit. The hotel chain pinned a promotion about its Private Jet Experience in Bora Bora, generating the following results:

- 8,564 re-pins and 9,121 visits to the brand's landing page of Around the World on a Private Jet.
- 463 views of 'Reservation Inquiry'.
- 483 clicks on the 'Request a Brochure' button, and 112 people requested a brochure for a trip that would cost US$70,000 per person.

It became obvious that the general public was not Four Seasons' target customers for products with such a high price tag. Nevertheless, the above responses generated from a pin appeared to be significant, although the company's senior vice president of marketing and communications also admitted that it was difficult to prove a direct connection between a pin and an actual purchase or booking.

Restaurant examples

Many of us probably still keep a fresh memory of 'cronut', a croissant-doughnut hybrid. When it got started, my friends visiting the States from overseas would request me to treat them with cronuts because they heard about this new creation and saw the pictures of cronuts online. The viral food moment of cronut began on May 10, 2013, according to Morell (2013). Dominique Ansel, the creator of cronut, launched his new product in his bakery shop in Soho but did not pay much effort in promoting his new creation. At the end of the day, he found a small write-up about cronut in an online foodie magazine. That was it. The next day when he woke up, he surprisingly found 140,000 links to photos (visual content) of cronuts on the internet. Moreover, people began lining up for the cronut even before the bakery opened its door. Since then, lines outside the bakery began to form at 5:30 am every day.

Because Dominique only made 200 – 250 cronuts a day, not everyone waiting in line was able to try the new creation, even though the shop had enforced a two-cronuts-per-customer policy. Some disappointed customers broke down crying and raised their voices at the staff. There was even a black market for cronuts, asking for up to US$40 a piece on Craigslist. They only cost US$5 each if purchased in the store. In the end, cronuts became one of the most Instagrammed dishes in New York and made the 2014 TIME list of "Objects that Move the News" (Sportelli, 2016). Bakeries around the globe started copying the concept.

Also in New York City's SoHo neighborhood, the chef-and-owner of Black Tap Craft Burgers & Beer, Joe Isidori, created signature milkshakes and posted a couple of pictures of the shakes online, reported by Sportelli (2016). In a few days, Joe began seeing consumer groups coming in for one of the signature milkshakes. They took pictures of the shakes with their phones. A few months later, Buzzfeed featured a story about the milkshakes. Soon, the story reached 1.6 million page views.

Through the publicity of relevant visual and textual content on the internet (i.e., pictures on Instagram and featured articles on a website), people began lining up for the restaurant at 10:00 am for its opening at 11:30 am. The lineup remained until about midnight. The 15-seat restaurant served 350 – 400 people a day.

Joe believed that good content on social media helped him boost his business. Every milkshake photo on the restaurant's Instagram page received more than 10,000 likes and thousands of comments. When the restaurant asked the customers, who were waiting in line, the response had been consistent – they read about the sakes online and wanted to share a photo with their followers once they got one.

The example at Airbnb

Marissa Phillips (n.d.), Head of Content Strategy at Airbnb, describes on the company's website that content strategists play 'an integral role' in shaping Airbnb's core product experiences. The content strategist team at Airbnb crafts messages for various audiences in a consistent voice – "*straightforward, inclusive, thoughtful, and spirited.*" Their role in the organization is well-defined, including:

- Get involved early, even in the product development process.
- Work with the designers to weigh different approaches to the design and content, and form recommendations through research and data science.
- Work with product managers to understand the issues and the information needed to convey people to use their products.
- Collaborate with the engineers, legal team, and language managers on its localization team to make sure their ideas translate well.
- Keep control of all Airbnb's words/contents shared in the cyber marketplace.
- Help Airbnb hosts in writing a description of their homes, prices for similar listings in the neighborhood, ideas around what Airbnb guests look for, and other useful tips.
- Communicate with the guests who booked their first Airbnb via e-mail, telling them what to expect and what is unique about Airbnb.

It is difficult to measure how Airbnb's content strategies contribute to the company's overall success. Nevertheless, it is clear that Airbnb managed to bounce back in the summer of 2020, even when the COVID-19 pandemic made an unprecedented impact on the hospitality and tourism industry (Kwok, 2020). For example:

- During the weekend of June 5 to June 7, Airbnb's gross booking recorded a year-to-year growth for the first time since February.
- Airbnb attracted over 100,000 new guests to use its short-term residential rental service.
- Airbnb observed a 22% year-to-year increase in booking in July.
- People book long-term stays at Airbnb, usually for weeks or months.
- Airbnb filed an initial public offering (IPO) in 2020.

Helpful tactics in content marketing

Companies can publish almost any content they want on the internet. They may also choose to share an update in the platforms/channels that they deem most effective in reaching their target customers. Not all content will receive the same attention from internet users, however. Research has shown that different types of content in a company's social media message would have various impacts on internet users' reactions toward the message, such as Likes and Shares on Facebook, Twitter, and Instagram (e.g., Kwok & Yu, 2013 & 2016; Leung *et al.*, 2017; Sabate *et al.*, 2014). Then, because internet users' reactions on social media websites often appear in their network's feeds, a more significant eWOM (electronic word-of-mouth) effect can be created (Kwok *et al.*, 2019). Hospitality and tourism businesses must be very selective in the types of content they publish and share if they want their messages to be heard or seen by a large number of internet users. Based on a narrative review of relevant literature, they are highly encouraged to incorporate the following research findings into their content marketing strategies.

Develop the right content

Kwok and Yu (2013) were probably among the first group of researchers who assessed internet users' reactions toward different content shared on social networking sites. They discovered in their analysis of 982 Facebook messages initiated by ten restaurant chains and two independent restaurants that businesses may publish two distinguishable content typed on social media websites. On the one hand, a restaurant may share content that directly sells or promotes their business, so-called 'sales/marketing messages'. A typical example of a marketing message can be: "Hey, Chili's fans! We are giving you a special sneak peek at one of our holiday offers! Starting today, you can receive 10 percent off any purchase of Chili's gift cards totaling US$100 or more! Only you, our Facebook fans know about this today! (followed by a URL)." On the other hand, a restaurant may share content without directly promoting a brand, a product, or the company, which they called 'conversational messages'. A conversational message may read: "Hey, Chili's fans! We hope everyone has a Happy Thanksgiving spent with family and friends. We'd love to hear your Turkey Day stories!" Facebook users tend to pay more attention and reacted better to conversational messages over sales/marketing messages if measured in the number of likes and comments that a message received.

Later, Kwok and Yu (2016) further examined 2,654 Facebook messages initiated by ten restaurant chains, two independent restaurants, 12 hotel chains, and two casinos. They proposed and advanced five sub-categories of the sales/marketing messages and four sub-categories of the conversational messages. When promoting their brand or product, hospitality companies may share content about their support or commitment to a community or a global society (Social Responsibility), boast about their businesses, brands, or products directly or indirectly (Direct Boasting and Indirect Boasting), highlight their products or services (Product Highlight), or run a sales/promotion campaign (Campaign/Promotion). When engaging their Facebook followers in a conversation, hospitality companies can call for a special action (Call for Action), such as asking fans to follow them on Twitter and other social media accounts, ask their fans questions to provoke feedback (Provoke Feedback), offer useful or helpful information to Facebook users (Advice/Suggestions), and inform Facebook users what is happening (Updates). Their analysis revealed that Facebook users gave more likes to the messages of Direct Boasting, Product Highlight, Call for Action, Provoke Feedback, Advice/Suggestions, and Updates. Facebook users also posted more comments on Provoke Feedback and Call for Action messages than other content.

Leung *et al.* (2017) examined 1,837 business-initiated messages from 12 hotel brands' Facebook pages in a similar study. They classified the messages into six categories: involvement (e.g., asking for feedback), product (e.g., what is offered in a hotel), promotion (e.g., special offers and discounts), brand (e.g., honor and awards received, hotel news, charity contributions), reward (i.e., games or contests that allow Facebook users to win rewards with or without a purchase), and information (e.g., holiday greetings and food or consumer trends in general). They found that product, brand, and involvement messages were more effective than the other three message types in getting more likes, comments, and/or shares among Facebook users.

The above research evidence indicates that the content type of a company's social media messages plays a significant role in influencing internet users' reactions toward the messages. Hospitality and tourism companies are strongly recommended to carefully select the right content as they develop an effective content marketing strategy.

Present the content in the right medium

According to the media richness theory, media fitness will affect communication efficiency among people (Daft & Lengel, 1986). Different media carries various ability in influencing people's understanding of the communicators (Kahai & Cooper, 2003). The function of media richness is defined as the medium's ability to carry multiple information cues, facilitate feedback, establish personal focuses, and utilize natural language (Armengol *et al.*, 2017). Because humans process visual information 60,000 times faster than text, visual content will receive significantly greater attention from internet users (Vilnai-Yavetz & Tifferet, 2015). Moreover, visual information revealed in pictures or videos can be more straightforward to most internet users than textual information because people with various racial or educational backgrounds no longer need to overcome the language barriers to interpret the information presented in the content. As a result, the medium choice also plays a significant role in influencing a content marketing strategy's effectiveness.

A large number of empirical studies have confirmed visual content's significant impact on internet users' attention to business-initiated messages. For example, Kwok and Yu (2013) and Leung *et al.* (2017) reported in their investigations that Facebook updates with photos, as compared to those with pure texts, a video, or a hyperlink, would receive more likes and comments. Sabate *et al.* (2014) also confirmed photos as an effective medium for businesses to increase brand popularity on Facebook. In another setting, tweets with photos can also help businesses substantially improve their customer engagement on Twitter (Li and Xie, 2020).

Besides photos, videos also play an increasingly important role in getting social media users' attention. Kwok *et al.* (2015) studied restaurants' Facebook posts over three different periods (fourth quarter of 2010, 2012, and 2014). They discovered that restaurants seemed to share more videos as time went by since 2010. Moreover, they discovered that videos had emerged in 2012 and 2014 as another essential medium to attract internet users' attention, showing no significant statistical difference from photos.

Back in 2010, Facebook messages with photos received a significantly higher number of likes and comments than videos.

Overall, it appears that visual content in photos and videos can gain significantly greater attention from internet users than purely textual information or message with hyperlinks. It is not surprising to see many social media platforms make most updates look like photos now. Facebook and Twitter, for example, now display a preview picture of the posts with a hyperlink and a video, using a photo from the website where the hyperlink direct or a screenshot of the video. Moreover, Facebook also adds a photo frame and visual background to the short textual updates, giving users the option of turning their posts with pure text into visual images. There is also surging popularity among the apps for visual content sharing, including TikTok, Instagram, and Pinterest, leading to the next topic of selecting the right distribution channel for content marketing.

Distribute the right content through the right channels

A few years ago, Facebook was the world's most visited website and the place where Americans spent the most time (Kwok & Yu, 2013). Facebook became the most popular social media website for B2C communications among organizations (Kwok & Yu, 2016). Accordingly, many companies put Facebook as the No. 1 channel for content distribution. There were, however, several backlashes toward Facebook's data privacy issues and its CEO (Fiegerman, 2018). The Federal Trade Commission (2020) also filed an antitrust lawsuit against the company in December 2020. Facebook is no longer the cool cyber place where everybody hangs out. Some internet users also called for a #DeleteFacebook movement (Binder, 2019). Then, when people are leaving Facebook, should hospitality and tourism organizations also shift their content distribution channel to other places? More importantly, through which (new) channels should hospitality and tourism organizations communicate with their prospective and existing customers? The answers depend on who is the organization's target customer.

YouTube and Facebook are the most popular channels for U.S. adults

Among U.S. adults, Facebook remained to be one of the most-used social media platforms in 2019. According to a *Pew Research Report* by Perrin and Anderson (2019), 69% of U.S. adults still used Facebook, a little bit behind YouTube at 73%. Other most-used social media platforms include Instagram (37%), Pinterest (28%), LinkedIn (27%), Snapchat (24%), and Twitter (22%). Particularly, Instagram and Snapchat, the platforms for visual content distribution, were well-received among young adults. Among 18-to-24-years old, 75% used Instagram, and 73% used Snapchat. Among 25-to-29-years old, 57% and 47% used Instagram and Snapchat, respectively.

Perrin and Anderson (2019) further reported that more males used YouTube than females (78% vs. 68%), but more females used Facebook (75% vs. 63%), Instagram (43% vs. 31%), and Pinterest (42% vs. 15%). A larger percentage of Hispanics used Instagram, at 51%, compared to African Americans (40%) and Caucasian whites (33%). LinkedIn was more popular among users with a college degree or above (51%) than those who had some college (26%) or a high school diploma or lower (9%). Largely because there is a strong correlation between people's educational attainment and their household incomes, LinkedIn was more adopted by the people with a household income of US$75,000 or above (41%), whereas only 27% of those making between US$30,000 and $74,999 and 18% of those making below-US$30,000 used the platform. LinkedIn has become the most preferred social networking site for job search and recruitment/selection purposes (Kwok & Muñiz, 2021).

YouTube, Instagram, and Snapchat are more popular than Facebook among U.S. teens

Anderson and Jiang (2018) provided the details about U.S. teens' social media usage in another *Pew Research Report.* In 2018, only 51% of teens used Facebook, significantly lower than YouTube (85%), Instagram (72%), and Snapchat (69%). Moreover, when they were asked which social media platform they used most often, only 10% mentioned Facebook, behind Snapchat (35%), YouTube (32%), and Instagram (15%). Three years ago, 71% of U.S. teens used Facebook, and Facebook was the most often used social networking site at 41% (Lenhart, 2015). According to another Statista report (Clement, 2020), the percentage of U.S. teens using Facebook dropped from 42% in 2012 to merely 2% in 2020. As of 2020, U.S. teams reported that their favorite social networking sites included Snapchat (34%), TikTok (29%), Instagram (25%), Twitter (3%), Discord (3%), and Facebook (2%). Also reported by Anderson and Jiang (2018), only 36% of the U.S. teens living in a household with a yearly income of US$75K or above used Facebook, and 70% of those teens in a household making less than US$30K a year used Facebook. Similar results about household incomes and Facebook usage were reported in another study with a different sample by the Associated Press – NORC Center for Public Affairs Research (2016) at the University of Chicago.

Facebook is still an essential channel in content marketing

It is worth mentioning that Facebook is the mother company of Instagram. The four most downloaded apps in the last decade, namely Facebook, Messenger, WhatsApp, and Instagram, were also owned by Facebook (Shead, 2019). More importantly, 74% of the adult Facebook users still checked the website every day, and 43% of U.S. adults got news on Facebook, the highest among all social media channels (Gramlich, 2019). Therefore, if a hospitality or tourism organization wants to announce or share an update with the public, Facebook can still be very effective. The right channel for content distribution must be carefully selected to tailor to the taste of a business' target customers.

Other considerations

The timing of content-sharing and message consistency are among other influential factors that affect the effectiveness of an integrated content marketing strategy (Batra & Keller, 2016). For example, it is more likely for a brand to get more attention from its target customers on social media websites during the time when they are active on the platforms. The good news is that many social media websites share user insights on a brand page with the business owners. It is not difficult for a content marketing manager to figure out when is the best time to share a certain type of content on the social media platform they manage. Another consideration is to share consistent brand messages over a long period of time and across different social media platforms. Message consistency helps reinforce the critical information being conveyed in a series of messages.

Conclusion

Content marketing is a relatively new concept and still evolving. While hospitality and tourism companies are highly encouraged to follow some general rules about content marketing (e.g., sharing more visual content than purely textual content), it is essential to test and try out different content marketing tactics in the social media platforms they managed. Customized content marketing strategies tailored to target customer groups on specific social media platforms may very likely work better than the one-size-fits-all strategies.

Discussion questions

1. How important is content marketing in today's business?
2. What content marketing strategies do hospitality and tourism companies usually use? How effective are they?
3. What other recommendations will you make to hospitality and tourism companies that want to implement a content marketing strategy?
4. What advice will you provide to people who want to start a career in content marketing?

References

American Marketing Association. (n.d.) *What is digital marketing?*, www.ama.org/pages/what-is-digital-marketing/ [Accessed: 7 January 2021].

Anderson, M. & Jiang J. (2018) *Teens, Social Media & Technology 2018*, Pew Research Center. www.pewresearch.org/internet/2018/05/31/teens-social-media-technology-2018/ [Accessed: 24 January 2020].

Armengol, X., Fernandez, V., Simo, P. & Sallan, J.M. (2017) An examination of the effects of self-regulatory focus on the perception of the media richness: The case of e-mail, *International Journal of Business Communication*, **54** (4), 394-407.

Associated Press – NORC Center for Public Affairs Research. (2016) *Instagram and Snapchat are most popular social networks for teens; Black teens are more active on social media, messaging apps*, The University of Chicago, http://apnorc.org/projects/pages/html%20reports/instagram-and-snapchat-are-most-popular-social-networks-for-teens.aspx [Accessed: 24 January 2020].

Bagley, E. & Lanz, L. (2014) The digital direction: Content management and the marketing communications of today (Part II of a two-part series), *HVS Global Hospitality Report*. **24**, 1-6.

Batra, R. & Keller, K.L. (2016) Integrating marketing communications: New findings, new lessons, and new ideas, *Journal of Marketing*, **80** (6), 122-145.

Binder, M. (2019) *Here's why you should finally #DeleteFacebook in 2020*, mashable.com/article/delete-facebook-2020/ [Accessed: 24 January 2020].

Clement, J. (2020) *Favorite social networks of U.S. teens 2012-2020*, Statista.com. www.statista.com/statistics/250172/social-network-usage-of-us-teens-and-young-adults/ [Accessed: 1 January 2020].

Content Marketing Institute. (2021) *What is digital marketing?*, https://contentmarketinginstitute.com/what-is-content-marketing/ [Accessed: 18 January 2021].

Daft, R.L. & Lengel, R.H. (1986) Organizational information requirements, media richness and structural design, *Management Science*, **32** (5), 554-571.

De Chernatoy, L. & Segal-Horn S. (2001) Building on services characteristics to develop successful service brands, *Journal of Marketing Management*, **17**, 645-669.

Federal Trade Commission (2020) *FTC sues Facebook for illegal monopolization*, https://www.ftc.gov/news-events/press-releases/2020/12/ftc-sues-facebook-illegal-monopolization [Accessed: 9 December 2020].

Fiegerman, S. (2018) *Facebook faces new regulatory backlash over data privacy*, CNN Business, https://money.cnn.com/2018/06/04/technology/facebook-data-backlash/index.html [Accessed: 24 January 2020].

Gramlich, J. (2019) *10 facts about Americans and Facebook*, Pew Research Center, https://www.pewresearch.org/fact-tank/2019/05/16/facts-about-americans-and-facebook/ [Accessed: 27 December 2020].

Järvinen, J. & Taiminen, H. (2016) Harnessing marketing automation for B2B content marketing, *Industrial Marketing Management*, **54**, 164-175.

Kahai, S.S. & Cooper, R.B. (2003) Exploring the core concepts of media richness theory: The impact of cue multiplicity and feedback immediacy on decision quality. *Journal of Management Information Systems*, **20** (1), 263-299.

Koiso-Kanttila, N. (2004) Digital content marketing: A literature synthesis, *Journal of Marketing Management*, **20**, 45-65.

Kwok, L. (2020) *Will more hotels get into the home-sharing business?*, HospitalityNet.org, www.hospitalitynet.org/opinion/4100791.html [Accessed: 24 January 2021].

Kwok, L. Mao, Z. & Huang, Y. K. (2019) Consumers' electronic word-of-mouth behavioral intentions on Facebook: Does message type have an effect?, *Tourism and Hospitality Research*, **19** (3), 296-307.

Kwok, L. & Muñiz, A. (2021) Do job seekers' social media profiles affect hospitality managers' hiring decisions? A qualitative inquiry, *Journal of Hospitality and Tourism Management*, **46**, 153-159.

Kwok, L. & Yu, B. (2013) Spreading social media messages on Facebook: An analysis of restaurant business-to-consumer communications, *Cornell Hospitality Quarterly*, **54** (1), 84-94.

Kwok, L. & Yu, B. (2016) Taxonomy of Facebook messages in business-to-consumer communications: What really works?, *Tourism and Hospitality Research*, **16** (4), 311-328.

Kwok, L., Zhang, F., Huang, Y.K., Yu, B., Maharabhushanam, P. & Rangan, K. (2015) Documenting business-to-consumer (B2C) communications on Facebook, *Worldwide Hospitality and Tourism Themes*, **7** (3), 283-294.

Lenhart, A. (2015) *Teens, social media & technology overview 2015*, Pew Research Center, https://www.pewresearch.org/internet/2015/04/09/teens-social-media-technology-2015/ [Accessed: 24 January 2020].

Leung, X., Bai, B. & Erdem, M. (2017) Hotel social media marketing: A study on message strategy and its effectiveness, *Journal of Hospitality and Tourism Technology*, **8** (2), 239-255.

Li, Y. & Xie, Y. (2020) Is a picture worth a thousand words? An empirical study of image content and social media engagement, *Journal of Marketing Research*, **57** (1), 1-19.

Morell, K. (2013) *Dominique Ansel: Meet the man behind the cronut craze*, AmericanExpress.com, https://www.americanexpress.com/en-us/business/trends-and-insights/articles/dominique-ansel-meet-the-man-behind-the-cronut-craze/ [Accessed: 24 January 2021].

Perrin, A. & Anderson, M. (2019) Shares of U.S. Adults using social media, including Facebook, is mostly unchanged since 2018, Pew Research Center, www.pewresearch.org/fact-tank/2019/04/10/share-of-u-s-adults-using-social-media-including-facebook-is-mostly-unchanged-since-2018/ [Accessed: 24 January 2020].

Phillips, M. (n.d.) *Content strategy at Airbnb – Building our practice*, Airbnb.com. https://airbnb.design/content-strategy-at-airbnb/ [Accessed: 24 January 2021].

Ramesh, M. & Vidhya, B. (2019) Digital marketing and its effect on online consumer buying behavior, *Journal of Services Research*, **19** (2), 61-77.

Rowley, J. (2008) Understanding digital content marketing, *Journal of Marketing Management,* **24** (5-6), 517-540.

Sabate, F., Berbegal-Mirabent, J., Cañabate, A. & Lebherz, P. R. (2014) Factors influencing popularity of branded content in Facebook fan pages, *European Management Journal,* **32** (6), 1001-1011.

Shankman, S. (2014) *Four Seasons' Pinterest strategy is focused on inspiration, not bookings,* https://skift.com/2014/05/02/four-seasons-pinterest-strategy-is-focused-on-inspiration-not-bookings/ [Accessed: 20 January 2021].

Shead, S. (2019) *Facebook owns the four most downloaded apps of the decade,* BBC News, https://www.bbc.com/news/technology-50838013 [Accessed: 1 January 2021].

Smith, D. (2016) *Make social media work for your hotel via storytelling,* HotelBusiness.com, https://assets.pixlee.com/website/hubspot/pdf/Case-Studies/Hotel-Business-Magazine-Pixlee.pdf [Accessed: 20 January 2021].

Sportelli, N. (2016) *Instagram is putting restaurants on the map, but what happens after their food goes viral?,* Forbes Magazine, https://www.forbes.com/sites/nataliesportelli/2016/03/24/instagram-is-putting-restaurants-on-the-map-but-what-happens-after-their-food-goes-viral/ [Accessed: 20 January 2021].

Steimle, J. (2014) *What is content marketing?,* Forbes Magazine, https://www.forbes.com/sites/joshsteimle/2014/09/19/what-is-content-marketing/ [Accessed: 18 January 2021].

Vilnai-Yavetz, I. & Tifferet, S. (2015) A picture is worth a thousand words: Segmenting consumers by Facebook profile images, *Journal of Interactive Marketing,* **32**, 53-69.

Wymbs, C. (2011) Digital marketing: The time for a new 'academic major' has arrived, *Journal of Marketing Education,* **33** (1), 93-106.

8 Examining contextual factors influencing hotel mobile app satisfaction

Liza Cobos

Introduction

The acceptance, adoption, and use of mobile applications allows businesses to reach new consumers and enhances the customer experience by properly using the features of mobile technology (Middleton *et al.*, 2014). Distinctive features of mobile technology such as portability, user identification, localization capability, and instant connectivity provide value to hospitality firms and consumers (Picoto *et al.*, 2014). Organizations that use and properly leverage mobile technologies are likely to gain an advantage and provide increase value and revenue to their business. Mobile business, such as mobile apps, provides firms with a unique value proposition by allowing easy, useful, and real-time access to information to consumers and personalization capabilities through client identification and localization services (Clarke, 2001; Watson *et al.*, 2002; Camponovo & Pigneur, 2003; Sharma & Gutiérrez, 2010; Picoto *et al.*, 2014).

The proliferation of new communication technology like mobile apps has changed the manner in which hospitality firms conduct and engage in business, and interact and communicate with consumers (Wang & Wang, 2010; Kansakar *et al.*, 2019). Such technologies have led to increased productivity and efficiency in service delivery by allowing firms to provide services to consumers through more convenient channels (Meuter *et al.*, 2003). Because of the benefits, consumers are insisting that firms provide new and innovative ways to conduct business (Xiang *et al.*, 2015; Kansakar *et al.*, 2019). Because of consumers' changing expectations, firms must become more consumer focused, flexible, innovative, and adaptable (Bilgihan *et al.*, 2011). Firms have typically responded to consumers' needs by dedicating additional resources to implementing new technologies. However, the branded mobile app has received limited attention in lodging research, leaving the industry with a lack of understanding on how to implement it as part of the firm's mobile commerce strategy (Kim & Law, 2015) and how the consumer is employing and interacting with branded mobile apps (Peng *et al.*, 2014).

Introducing contextual factors allows researchers to understand the impact of technology-specific characteristics to leverage their use when implementing said technology. To that end, the objective of the study is to examine the contextual and cognitive factors that may influence consumer satisfaction with a hotel mobile app. Specifically, the author wants to determine which factor is the best predictor to satisfaction.

Literature review

Technology research in the hospitality industry has been primarily dedicated on consumer acceptance and adoption of technology specifically looking at the influence of motivators, inhibitors, and perceptions (Law *et al.*, 2018). Attitude theories of technology adoption look at factors influencing a user's intention to adopt, such as usefulness, ease of use, performance expectancy, social influence, facilitating condition, and effect expectancy (Davis, 1989; Venkatesh *et al.*, 2003). Research regarding mobile applications in the hospitality and tourism industry has focused on the adoption of technologies with Davis' (1989) Technology Acceptance Model (TAM) being the most frequently cited theory to explore adoption (Law *et al.*, 2018). TAM has been used to study travel and lodging mobile apps (Im & Hancer, 2014; Rivera *et al.*, 2015).

Consumer behavior theories like Expectation-Confirmation Theory (Oliver, 1980) allow for researchers to study consumer satisfaction by examining the consumer's expectations and their pre- and post-purchase performance to measure satisfaction and its impact on repurchase intentions. This study will use the Expectation-Confirmation Theory as a foundation and include contextual factors to the model. These factors were added as they are an area that needs further exploration in order to improve the customer experience (Law *et al.*, 2018). Examining satisfaction allows for a better understanding of factors influencing the use of mobile apps. Satisfaction has consistently been of great interest to scholars as it has a strong impact on decision-making behavior (Zhang, 2013). Consequently, a better understanding of satisfaction will lead to practical implications for the design, acceptance, and management of new technologies like mobile apps (Zhang, 2013).

Contextual factors

Mobility

This study examines context-specific characteristics of personalization, mobility and responsiveness that may influence a consumer's satisfaction with a hotel mobile app. Mobility is what makes mobile technology unique and valuable for both the consumer and the firm. It allows firms to reach consumers and eases information search and decision-making (Mallat *et al.*, 2009). In this study, 'mobility' refers to the users' awareness of the mobility value of mobile apps (Huang *et al.*, 2007). The literature proposes that mobility has three elements: expediency, convenience, and immediacy (Seppälä & Alamäki, 2003). The mobility function of mobile devices allows users access to information, immediate communication, and availability to services anytime and anywhere (Mallat *et al.*, 2009). Perceived mobility has been described as a critical factor for mobile service users (Yen & Wu, 2016). Because of this characteristic, it can be expected that mobility is an essential factor of consumer mobile app adoption, use and satisfaction (Yen & Wu, 2016). Mobility has been found to be an important antecedent to the users' assessment of their satisfaction with mobile apps (Lu *et al.*, 2017). Satisfaction has also been examined as an influencing factor to mobility; it has been found that satisfaction exerts a considerable influence on perceived mobility in the context of mobile social networks (Park *et al.*, 2014). Due to the limited research on the aspect of mobility, the author posits the following:

Hypothesis 1: Perceived mobility of the hotel mobile app positively impacts users' satisfaction with a mobile app.

Personalization

Personalization in the information systems environment has been defined as modification of the functions, information, message, or interface with the goal of increasing the personal relevance to the user (Blom, 2000). Similarly, personalization has been described as system-driven, system-initiated and requires adaptive components (Treilblmaier *et al.*, 2004). In electronic and mobile commerce, personalization is a way to gather consumer data, on search and purchase behavior, to make product recommendations to consumers.

In order to achieve personalization, detailed information from the user is needed to make appropriate modifications for the user's needs. In addition, effective personalization requires the system to constantly monitor user behavior, which enables the system to adjust recommendations and/or messages based on the behavior. Personalized systems utilize user profiles that are continuously managed and updated by the system. The hospitality industry utilizes personalization to perform mobile commerce marketing activities like hotel email blasts for promotions based on the consumers' previous visits, or push notifications during a hotel stay for concierge and other hotel services. Research on mobile commerce suggests that personalization increases consumption by improving the fit between the consumers' needs and product features (Wattal *et al.*, 2009). Scholars explain that even though personalization is viewed as an essential factor of the present-day service experience, theory has not been able to converge toward unified findings that clearly establish personalization as an influencing factor of mobile commerce behaviors (Lee & Cranage, 2011; Sutanto *et al.*, 2013). Results from studies examining mobile services and technologies have found that personalization exerts a positive influence on satisfaction in online and mobile contexts like websites and social media apps (Lin *et al.*, 2012; Barnes & Vidgen, 2014; Park, 2014). The influence of personalization has also been examined as part of process quality factors and has been found to indirectly influence satisfaction in the context of mobile apps (Chen *et al.*, 2012). Given the importance of personalization for the successful implementation of mobile commerce to better service and influence consumer behavior (Dickinson *et al.*, 2014), the author proposes the following hypothesis:

Hypothesis 2: Perceived personalization of the hotel mobile app positively impacts users' satisfaction with the mobile app.

Responsiveness

The third contextual factor examined in this study relates to the mobile apps responsiveness. Responsiveness discusses the degree to which users can find information pertinent to their request (Burgoon *et al.*, 1999). Responsiveness is included since it is one of the main features of mobile technologies and a motivation for its adoption and use (Choi *et al.*, 2015; Yoo *et al.*, 2015). In addition, it has been identified as one of the biggest challenges the hospitality industry is facing related to the implementation of mobile technologies and mobile devices (Kansakar *et al.*, 2019).

Previous studies on e-WOM systems have examined responsiveness as one of the elements involved in the interactivity traits leading to satisfaction (Yoo *et al.*, 2015). Their results show that among the interactivity components, responsiveness is the most influential in the interactivity of e-WOM systems. Benlian *et al.*'s (2011) study of organizational software examined system quality using responsiveness as one of its components. Responsiveness was found to be the most significant element of system quality, which in turn was found to be an antecedent to perceived satisfaction. Responsiveness of mobile apps, as an element of system quality, has been found to indirectly influence user satisfaction (Chen *et al.*, 2012). Further review of the literature showed that responsiveness has been examined as one of the elements of e-service quality influencing the user's utilitarian value (Mouakket, 2014). However, responsiveness was not found to be a significant dimension of the utilitarian value among the users of online reservation systems. As discussed, previous studies have examined responsiveness as one of the factors of system quality or service quality. Since this study seeks to examine the influence of mobile technology attributes and their impact on satisfaction, the following hypothesis was drafted:

Hypothesis 3: Perceived responsiveness of the hotel mobile app positively impacts users' satisfaction with a mobile app.

Cognitive factors

In addition to contextual factors, this study sought to examine cognitive factors to predict satisfaction. Cognitive factors are important to consider in mobile technology research, as they are deemed important due to their influence on users' feelings about a technology (Holbrook, 1978). Cognitive factors represent the conscious decisions relating to the behavioral purpose of serving the user's needs. This study considered the cognitive factors of perceived usefulness of the mobile app and confirmation of expectations.

Perceived usefulness

Perceived usefulness is defined as a user's perception of the anticipated benefits of utilizing a technology and information systems (IS) (Davis *et al.*, 1989). Perceived usefulness is a commonly used construct in technology research and has been found to significantly influence technology use and adoption (Davis *et al.*, 1989; Kim *et al.*, 2008; Nabavi *et al.*, 2016). Furthermore, it is the only belief construct that has been found to constantly influence a user's intention across different stages of IS use (Bhattacherjee,

2001; Davis *et al.*, 1989; Karahanna *et al.*, 1999). Perceived usefulness has been found to influence satisfaction among augmented reality application users (Kim *et al.*, 2014); mobile instant messaging (Oghuma *et al.*, 2016); and online reservation system users (Mouakket, 2014). In the mobile travel booking service context, perceived usefulness was found to impact satisfaction (Zhong *et al.*, 2015). From this discussion, it is apparent that the impact of usefulness is important and must be examined. The following relationship was examined:

Hypothesis 4: Perceived usefulness of a hotel mobile app positively impacts users' satisfaction with the mobile app.

Confirmation

Confirmation is the users' perceptions of the comparison between the expectation of a hotel mobile app and its actual performance. This is a measure of the confirmation of expectations after actual use (Oliver, 1980). Confirmation of expectations was found to positively impact satisfaction with the mobile app (Hsu & Lin, 2015). Furthermore, confirmation has been found to positively influence satisfaction in the long-term use of a system (Bhattacherjee & Premkumar; 2004; Nabavi *et al.*, 2016). However, the impact of satisfaction is not as robust as has been postulated. Halilovic and Cicic (2013) found that confirmation indirectly impacts the IS software long-term use in organizational settings through satisfaction. However, among the variables tested in that study, confirmation was found to have the least influence with conditions of support exerting the most influence. The results of the study show that the influence of the confirmation of expectations may not be as significant as previously posited by scholars (i.e. Bhattacherjee, 2001). Given the interesting results regarding the level of influence confirmation of expectations has on satisfaction further research is warranted. The following relationship was examined.

Hypothesis 5: Confirmation of expectations of a hotel mobile app positively impacts users' satisfaction with the mobile app.

Methodology

A self-administered online questionnaire distributed through an online panel was utilized to collect the data of the study. Online surveys have been widely adopted by scholars of many disciplines and deemed appropriate for researching online behaviors (Ayeh *et al.*, 2013; Lin & Filieri, 2015; Hung & Law, 2011). Participants recruited were screened for minimum age requirements (only people 18 and older could participate) and their previous use of hotel mobile apps. In order to take the survey, participants needed to have previous experience using hotel branded mobile applications. A total of 550 questionnaires were collected; after cleaning the data set of 478 were retained for analysis.

The questionnaire items were modified from the established literature and instruments. To reflect the context of the study the items were reworded. A 7-point Likert Scale (strongly disagree =1 to strongly agree =7) was used to measure the scale items. Four items were used to measure perceived usefulness. Sample items include "*Using hotel branded mobile apps improves my performance in managing hotel related services*" and "*Using hotel branded mobile apps increases my productivity in managing hotel related services*" (Bhattacherjee, 2001; Davis *et al.*, 1989). The confirmation of expectations construct was measured using three items. Examples include: "*My experience with using hotel branded mobile apps was better than what I expected*" and "*The service level provided by hotel branded mobile apps was better than what I expected*" (Bhattacherjee, 2001). Four items were used to measure perceived mobility including "*I know that hotel branded mobile apps are an instrument for hotel information and booking*" and "*It is convenient to access hotel branded mobile apps anywhere at anytime*" (Huang *et al.*, 2007; Yen & Wu, 2016). Perceived personalization was measured using three items. Examples of questions include: "*Hotel branded mobile apps can provide me with personalized services, products, or information tailored to my specific hotel stay*" and "*Hotel branded mobile apps can provide me with more relevant services, products, or information tailored to my preferences or personal interests*" (Li *et al.*, 2011:139). Four items were used to measure perceived responsiveness including "*Hotel branded mobile apps provide appropriate information to fulfill my needs*" and "*Hotel branded mobile apps provide the information that I need to fulfill my needs*" (Yoo et al., 2015). Lastly, satisfaction was measured using four items. Example include: "*I am satisfied with my overall experience with the hotel branded mobile app*" and "*I am pleased with my overall experience with the hotel branded mobile app*" (Bhattacherjee, 2001).

The profile of the respondents was similar between females (54.2%) and males (45.8%). More than half of the respondents reported that they were married (56.1%), with the second largest group reporting they were never married (34.1%). The remaining participants were divorced, separated, or widowed. In terms of mobile app experience, almost all respondents have been using mobile apps between 1 to 8 years (80%). Furthermore, almost 100% of the respondents reported using a hotel branded mobile app between 1 to 6 times per year (Table 8.1).

Table 8.1: Demographic characteristics of the participants

Variables	n	Percentage (%)
Gender		
Female	259	54.2
Male	219	45.8
Age		
18-24	39	8.2
25-34	199	41.6
35-44	137	28.7
45-54	68	14.2
55-64	24	5
65-74	11	2.3
Marital status		
Married/domestic partner	268	56.1
Never married	163	34.1
Divorced	31	6.5
Separated	9	1.9
Widowed	7	1.5
Education		
Some high school	2	.4
High school graduate	46	9.6
Some college	105	22
2 year college	60	12.6
4 year college	198	41.4
Professional degree	59	12.3
Doctorate	8	1.7

N= 478

Results

Data was analyzed using SPSS 24, which involved multiple stages. First multivariate assumptions of normality, linearity, homoscedasticity, and independence were verified, multicollinearity was not present in the data (Hair *et al.*, 2010). Reliability for each construct was tested; Cronbach Alpha coefficients exceeded the lower recommended limit of .70 (Hair *et al.*, 2010), coefficients ranged from .82 to .90. Table 8.2 shows the reliability for each construct.

Correlation coefficients were analyzed to assess collinearity, correlations for independent variables between .70 and .90 and .40 and .70, are considered high and moderate respectively (Hair *et al.*, 2007). In addition, variance inflation factor (VIF) values below 10 indicate that multicollinearity is not an issue (Hair *et al.*, 2010). Descriptive statistics and correlations are presented in Table 8.3.

Table 8.2: Conbrach's Alpha for study construct

Construct	α
Perceived usefulness	.896
Confirmation	.821
Perceived mobility	.886
Perceived personalization	.859
Perceived responsiveness	.875
Satisfaction	.901

Table 8.3: Descriptive statistics and correlations

		Mean	SD	1	2	3	4	5	6	VIF
1	Satisfaction	5.60	1.05	1.000						
2	Mobility	5.85	1.00	0.727*	1.000					3.614
3	Personalization	5.48	1.03	0.741*	0.748*	1.000				3.147
4	Responsiveness	5.73	0.95	0.751*	0.814*	0.766*	1.000			3.871
5	Usefulness	5.47	1.06	0.714*	0.743*	0.722*	0.755*	1.000		3.031
6	Confirmation of expectations	5.29	1.08	0.726*	0.671*	0.703*	0.668*	0.698*	1.000	2.400

*p<.01

To develop a model, Multiple linear regression analysis was used to predict customer satisfaction from usefulness, confirmation of expectations, perceived mobility, personalization and responsiveness. The results of the regression showed that the model explained 68.9% of the variance and that the model is a significant predictor of customer satisfaction, F(5, 472)= 209.367, p<.001. All predictor variables had a significant or high statistical significance on customer satisfaction, confirmation of expectations (B= .267, p <.001), responsiveness (B= .231, p <.001), personalization (B= .201, p <.001), mobility (B= .123, p<.05), and usefulness (B= .117, p<.05). The results lend support to all the hypotheses proposed in the current study.

Discussion

Technology plays a crucial role in many hospitality firms to engage, connect and facilitate user transactions. The current study confirmed the influence of cognitive and contextual factors on hotel mobile app user satisfaction. However, the results of the study show the effect of each factor is different. The results indicate that confirmation of expectations exerts the most impact, which mirrors results from previous studies (Hsu & Lin,

2015, Nabavi *et al.*, 2016). The results demonstrate that user expectations must be monitored and managed. Service providers should anticipate that user expectations will change over time. In order to achieve satisfaction, updates and improvements must be continuously made to the mobile application.

Results show that contextual factors deserve further evaluation, demonstrating the importance of studying the apps' characteristics. The results of the statistical analysis show that responsiveness, personalization and mobility exert a considerable influence on user satisfaction. The rapid changes in technology and its capabilities have been demonstrated to be a challenge for many firms when implementing new technology (Kansakar *et al.*, 2019). To keep up with the pace of changes and effective implementation of apps, it is imperative to understand the contextual factors of the technology. Studies that have examined responsiveness have found that it can indirectly influence satisfaction as a part of the measure of system quality and interactivity in e-WOM and organizational software (Benlian *et al.*, 2011; Chen *et al.*, 2012). IT scholars should continue to examine responsiveness as an independent measure and not as part of system quality or interactivity. The results suggest that having a mobile app that is able to provide information relevant to the user's request is integral to satisfaction.

The results of the study lend support to the posited influence of app personalization on satisfaction. Results mirror those from previous studies in which personalization in social media and websites has a positive influence satisfaction (Barnes & Vidgen, 2014; Lin *et al.*, 2012; Park, 2014). Increased personalization cannot only influence satisfaction but can also facilitate transactions and increase consumption (Wattal *et al.*, 2009). This further highlights the importance of the results, for most organizations are always looking for ways to increase and improve sales. This goal can be facilitated by increased consumption by current customers.

The results lend support to the impact of mobility on satisfaction. The findings are similar to those from other studies that have found that user awareness of the mobility functions of apps influence satisfaction with mobile apps (Lu *et al.*, 2017). Mobility is a feature that many users may take for granted since mobile technology has been the norm for many years. Service providers should place efforts to highlight the capabilities and functions that a user can perform 'on the go' such as check-in, keyless entry, and reservations, among others. Emphasizing all the functions that can be performed with the mobile app could lead to increased awareness and satisfaction with the app.

Lastly, the results support previous findings on the influence of usefulness on satisfaction. Perceived usefulness is a commonly used construct in technology research with results showing a significant influence on technology adoption and use (Davis *et al.*, 1989; Karahanna *et al.*, 1999). The findings are similar to previous studies examining technology in mobile instant messaging (Oghuma *et al.*, 2016), online reservations (Mouakket, 2014) and online travel booking (Zhong *et al.*, 2015). However, the results show that even though it had an influence, it was the least influential of the factors examined. The results are interesting; however, the results further demonstrate the need to continue to examine different factors and their impact on user satisfaction.

Conclusion

Mobile apps have been embraced by the hotel industry but scholars need to continue to expand the body of knowledge, given the complexities of the technology, to ensure successful implementation and management of mobile apps. Expanding research based on the technology's characteristics can help scholars identify aspects for more effective app design that is likely to influence consumer behaviors.

Information system scholars have called for the inclusion of contextual factors to add to theory in a systematic manner; embracing this approach allows for practical recommendations and richer theories (Hong *et al.*, 2014; Weber, 2003). Factors related to mobile technology, specifically mobile apps, help provide a better understanding of the influence those factors have on consumer behavior and emphasizes the relative importance of each factor. This allows practitioners and hospitality scholars to understand the importance of those factors and their influence on user satisfaction with the mobile app.

This study highlighted the importance of contextual factors for more effective mobile app implementation and customer satisfaction. Usefulness has been found to be the most influential factor in satisfaction in the context of technology (Davis *et al.*, 1989; Karahanna *et al.*, 1999; Nabavi *et al.*, 2016). However, the findings of this study suggest that contextual factors play a bigger role than previously imagined. The findings emphasize the importance of understanding the technology, its unique characteristics and the potential those characteristics have to influence satisfaction. Mobile apps are commonly used by many hotel brands, but scholars must strive to continue to expand the body of knowledge to understand the unique characteristics of the technology. Better understanding of those characteristics and their impact on the user can help organizations extract the most value from the technology implementation.

A number of practical implications that can be derived from this study. The amount of resources being spent by firms to implement mobile applications emphasizes the need to determine the factors that assist with customer satisfaction with mobile apps. Effective implementation of mobile apps can lead to satisfaction with the app and can have a positive influence on the customer experience. The results of the study suggest the apps' capability to personalize the experience, being responsive to the customer needs and wants, and the mobility features all have an impact on satisfaction. In addition, having the app meet the customer's expectations seems to hold the most important role on customer satisfaction with the mobile app.

The results provide additional understanding of customer use of hotel branded mobile apps and some factors that may lead to satisfaction with the app. This study helps highlight the importance of contextual factors for the successful deployment, use and management of mobile apps. Satisfaction with the app allows the firm to engage better with the customer, which may lead to use that is more consistent. Increased and consistent app use by the customer allows the firm to capture user data and provide personalized content. The information captured allows messages to be personalized and to cater to the needs of the customer, based on their search behavior and preferences.

No research is exempt from limitations. This research only considered a sample of mobile app users from the United States. Another limitation of this study is that results are not generalizable to non-branded apps. That is, the study examined the satisfaction with hotel branded mobile apps. Therefore, it does not include other non-branded or other types of apps available to the customer in other segments of the hospitality industry. Future research can examine other contextual factors to develop a deeper understanding the effect contextual factors have on satisfaction. This study examined the satisfaction with hotel branded mobile apps; future studies could examine the satisfaction with apps in other segments of the hospitality industry like museums, restaurants, food delivery, car rentals, airlines, parks and recreation. Finally, future research could examine app usage in different countries. This will help global brands examine if there are differences in factors affecting satisfaction with the mobile app.

Discussion questions

1. What other cognitive factors would help explain satisfaction in the context of hotel branded mobile apps?
2. Why is it important to study contextual factors in mobile technology research?
3. What other contextual factors should be examined in the mobile app context?

References

Ayeh, J.K., Au, N. & Law, R. (2013) Predicting the intention to use consumer-generated media for travel planning, *Tourism Management*, **35**, 132-143.

Barnes, S.J. & Vidgen, R.T. (2014) Technology socialness and web site satisfaction, *Technological Forecasting and Social Change*, **89**, 12-25.

Benlian, A., Koufaris, M. & Hess, T. (2011) Service quality in software-as-a-service: Developing the SaaS-Qual measure and examining its role in usage continuance, *Journal of Management Information Systems*, **28** (3), 85-126.

Bhattacherjee, A. (2001) Understanding information systems continuance: An expectation-confirmation model, *MIS Quarterly*, **25** (3), 351-370.

Bhattacherjee, A. & Premkumar, G. (2004) Understanding changes in belief and attitude toward information technology usage: A theoretical model and longitudinal test, *MIS Quarterly*, **28** (2), 229-254.

Bilgihan, A., Okumus, F. & Kwun, D.J.W. (2011) Information technology applications and competitive advantage in hotel companies, *Journal of Hospitality and Tourism Technology*, **2** (2), 139-154.

Blom, J. (2000) Personalization: A taxonomy, *Extended abstracts of the CHI 2000 Conference on Human Factors in Computing Systems*, 313-314.

Burgoon, J.K., Bonito, J.A., Bengtsson, B., Ramirez Jr, A., Dunbar, N.E. & Miczo, N. (1999) Testing the interactivity model: Communication processes, partner assessments, and the quality of collaborative work, *Journal of Management Information Systems*, **16** (3), 33-56.

Camponovo, G. & Pigneur, Y. (2003) Analyzing the m-business landscape, *Annales des Telecommunications*, **58** (1), 59-77.

Chen, L., Meservy, T.O. & Gillenson, M. (2012) Understanding information systems continuance for information-oriented mobile applications, *Communications of the Association for Information Systems*, **30** (1), 127-146.

Choi, M., Han, K. & Choi, J. (2015) The effects of product attributes and service quality of transportation card solutions on service user's continuance and word-of-mouth intention , *Service Business*, **9** (3), 463-490.

Clarke, I., (2001) Emerging value propositions for m-commerce, *Journal of Business Strategies*, **18** (2), 133-148.

Davis, F.D. (1989) Perceived usefulness, perceived ease of use, and user acceptance of information technology, *MIS Quarterly*, **13** (3), 319-340.

Davis, F.D., Bagozzi, R.P. & Warshaw, P.R. (1989) User acceptance of computer technology: A comparison of two theoretical models, *Management Science*, **35** (8), 982-1003.

Dickinson, J.E., Ghali, K., Cherrett, T., Speed, C., Davies, N. & Norgate, S. (2014) Tourism and the smartphone app: Capabilities, emerging practice and scope in the travel domain, *Current Issues in Tourism*, **17** (1), 84-101.

Hair, J.F., Black, W.C., Babin, B. J., & Anderson, R.E. (2010) *Multivariate Data Analysis*, 7th edn, New Jersey: Upper Saddle River Prentice Hall.

Hair, J.F., Money, A.H., Samouel, P., & Page, M. (2007) *Research Methods for Business*, New York: John Wiley & Sons, Inc.

Halilovic, S. & Cicic, M. (2013) Antecedents of information systems user behaviour–extended expectation-confirmation model, *Behaviour & Information Technology*, **32** (4), 359-370.

Holbrook, M.B. (1978) Beyond attitude structure: Toward the informational determinants of attitude, *Journal of Marketing Research*, **15** (4), 545-556.

Hong, W., Chan, F.K., Thong, J.Y., Chasalow, L.C. & Dhillon, G. (2014) A framework and guidelines for context-specific theorizing in information systems research, *Information Systems Research*, **25** (1), 111-136.

Hsu, C.L. & Lin, J.C.C. (2015) What drives purchase intention for paid mobile apps?–An expectation confirmation model with perceived value, *Electronic Commerce Research and Applications*, **14** (1), 46-57.

Huang, J.H., Lin, Y.R. & Chuang, S.T. (2007) Elucidating user behavior of mobile learning, *Electronic Library*, **25** (5), 585-598.

Hung, K. & Law, R. (2011) An overview of Internet-based surveys in hospitality and tourism journals, *Tourism Management*, **32** (4), 717-724.

Im, J.Y. & Hancer, M. (2014) Shaping travelers' attitude toward travel mobile applications, *Journal of Hospitality and Tourism Technology*, **5** (2), 177-193.

Kansakar, P., Munir, A. & Shabani, N. (2019) Technology in the hospitality industry: Prospects and challenges, *IEEE Consumer Electronics Magazine*, **8** (3), 60-65.

Karahanna, E., Straub, D.W. & Chervany, N.L. (1999) Information technology adoption across time: A cross-sectional comparison of pre-adoption and post-adoption beliefs, *MIS Quarterly*, **23** (2), 183-213.

Kim, D.J., Ferrin, D.L. & Rao, H.R. (2008) A trust-based consumer decision-making model in electronic commerce: The role of trust, perceived risk, and their antecedents, *Decision Support Systems*, **44** (2), 544-564.

Kim, H.H. & Law, R. (2015) Smartphones in tourism and hospitality marketing: A literature review, *Journal of Travel & Tourism Marketing*, **32** (6), 692-711.

Kim, K., Hwang, J., Zo, H. & Lee, H. (2014) Understanding users' continuance intention toward smartphone augmented reality applications, *Information Development*, **32** (2), 161-174.

Law, R., Chan, I.C.C. & Wang, L. (2018) A comprehensive review of mobile technology use in hospitality and tourism, *Journal of Hospitality Marketing & Management*, **27** (6), 626-648.

Lee, C.H. & Cranage, D.A. (2011) Personalisation–privacy paradox: The effects of personalisation and privacy assurance on customer responses to travel web sites, *Tourism Management*, **32** (5), 987-994.

Li, H., Sarathy, R. & Xu, H. (2011) The role of affect and cognition on online consumers' decision to disclose personal information to unfamiliar online vendors, *Decision Support Systems*, **51** (3), 434-445.

Lin, T.C., Wu, S., Hsu, J.S.C. & Chou, Y.C. (2012) The integration of value-based adoption and expectation–confirmation models: An example of IPTV continuance intention, *Decision Support Systems*, **54** (1), 63-75.

Lin, Z. & Filieri, R. (2015) Airline passengers' continuance intention towards online check-in services: The role of personal innovativeness and subjective knowledge, *Transportation Research Part E: Logistics and Transportation Review*, **81**, 158-168.

Lu, J., Liu, C. & Wei, J. (2017) How important are enjoyment and mobility for mobile applications? *Journal of Computer Information Systems*, **57** (1), 1-12.

Mallat, N., Rossi, M., Tuunainen, V.K. & Öörni, A. (2009) The impact of use context on mobile services acceptance: The case of mobile ticketing, *Information & Management*, **46** (3), 190-195.

Meuter, M.L., Ostrom, A.L., Bitner, M.J. & Roundtree, R. (2003) The influence of technology anxiety on consumer use and experiences with self-service technologies, *Journal of Business Research*, **56** (11), 899-906.

Middleton, C., Scheepers, R. & Tuunainen, V.K. (2014) 'When mobile is the norm: Researching mobile information systems and mobility as post-adoption phenomena', *European Journal of Information Systems*, **23** (5), 503-512.

Mouakket, S. (2014) Investigating the motives of customers' continuance intentions towards online reservation: The role of e-service quality, *Journal of Organizational and End User Computing*, **26** (2), 18-40.

Nabavi, A., Taghavi-Fard, M.T., Hanafizadeh, P. & Taghva, M.R. (2016) Information technology continuance intention: A systematic literature review, *International Journal of E-Business Research*, **12** (1), 58-95.

Oghuma, A.P., Libaque-Saenz, C.F., Wong, S.F. & Chang, Y. (2016) An expectation-confirmation model of continuance intention to use mobile instant messaging, *Telematics and Informatics*, **33** (1), 34-47.

Oliver, R.L. (1980) A cognitive model of the antecedents and consequences of satisfaction decisions, *Journal of Marketing Research*, **17** (4), 460-469.

Park, E., Baek, S., Ohm, J. & Chang, H.J. (2014) Determinants of player acceptance of mobile social network games: An application of extended technology acceptance model, *Telematics and Informatics*, **31** (1), 3-15.

Park, J.H. (2014) The effects of personalization on user continuance in social networking sites, *Information Processing & Management*, **50** (3), 462-475.

Peng, K.F., Chen, Y. & Wen, K.W. (2014) Brand relationship, consumption values and branded app adoption, *Industrial Management & Data Systems*, **114** (8), 1131-1143.

Picoto, W. N., Bélanger, F. and Palma-dos-Reis, A. (2014). 'An organizational perspective on mbusiness: usage factors and value determination', European Journal of Information Systems, 23(5), 571-592.

Rivera, M., Gregory, A. & Cobos, L. (2015) Mobile application for the timeshare industry: The influence of technology experience, usefulness, and attitude on behavioral intentions, *Journal of Hospitality and Tourism Technology*, **6** (3), 242-257.

Seppälä, P. & Alamäki, H. (2003) Mobile learning in teacher training, *Journal of Computer Assisted Learning*, **19** (3), 330-335.

Sharma, S. & Gutiérrez, J.A. (2010) An evaluation framework for viable business models for m-commerce in the information technology sector, *Electronic Markets*, **20** (1), 33-52.

Sutanto, J., Palme, E., Tan, C.H. & Phang, C.W. (2013) Addressing the personalization-privacy paradox: An empirical assessment from a field experiment on smartphone users, *MIS Quarterly*, **37** (4), 1141-1164.

Treiblmaier, H., Madlberger, M., Knotzer, N. & Pollach, I. (2004) Evaluating personalization and customization from an ethical point of view: An empirical study, *37th Annual Hawaii International Conference on System Sciences*, January, Los Alamitos, CA.

Venkatesh, V., Morris, M.G., Davis, G.B. & Davis, F.D. (2003) User acceptance of information technology: Toward a unified view, *MIS Quarterly*, **27** (3), 425-478.

Wang, H.Y. & Wang, S.H. (2010) Predicting mobile hotel reservation adoption: Insight from a perceived value standpoint, *International Journal of Hospitality Management*, **29** (4), 598-608.

Watson, R.T., Pitt, L.F., Berthon, P. & Zinkhan, G.M. (2002) U-commerce: expanding the universe of marketing, *Journal of the Academy of Marketing Science*, **30** (4), 333-347.

Wattal, S., Telang, R. & Mukhopadhyay, T. (2009) Information personalization in a two-dimensional product differentiation model, *Journal of Management Information Systems*, **26** (2), 69-95.

Weber, R. (2003) Editor's comments, *MIS Quarterly*, **27** (3), iii-xii.

Xiang, Z., Magnini, V.P. & Fesenmaier, D.R. (2015) Information technology and consumer behavior in travel and tourism: Insights from travel planning using the internet, *Journal of Retailing and Consumer Services*, **22**, 244-249.

Yen, Y.S. & Wu, F.S. (2016) Predicting the adoption of mobile financial services: The impacts of perceived mobility and personal habit, *Computers in Human Behavior*, **65**, 31-42.

Yoo, C.W., Kim, Y.J. & Sanders, G.L. (2015) The impact of interactivity of electronic word of mouth systems and E-Quality on decision support in the context of the e-marketplace, *Information & Management*, **52** (4), 496-505.

Zhang, P. (2013) The affective response model: A theoretical framework of affective concepts and their relationships in the ICT context, *MIS Quarterly*, **37** (1), 247-274.

Zhong, Z., Luo, J. & Zhang, M. (2015) Understanding antecedents of continuance intention in mobile travel booking service, *International Journal of Business and Management*, **10** (9), 156-162.

9 Big data analytics and database marketing

Guntekin Simsek and Sule Kaya Unal

Introduction

A data explosion has occurred worldwide with the connection of phones, computers, tablets, refrigerators, washing machines, watches, glasses, and many more devices to the internet (Mariani *et al.*, 2018: 3515). It has become almost meaningless to explain the level of usage of information technologies by consumers in the world. Since machines can even generate data on their own and users can use more than one technology simultaneously, it is impossible to determine the number of users accurately. Continuous knowledge flow has become an indispensable component in people's lives. Hence, technology that will support constant data flow, and business models that will store, analyze, and monetize data have been developed.

The change and development of work and life with technology have significantly affected consumer preferences. Adopting a more individualist, independent, and consumption-oriented approach, rather than focusing on production, has shaped postmodern consumers and their preferences. Thus, it is necessary to switch to personalized and unique applications rather than mass promotional activities to the new generation of consumers, who are more experienced, independent, and irrational (Inanc-Demir & Kozak, 2019: 220). Postmodern consumers' typical behaviors can be easily observed in the tourism activities they engage with and how they prefer to pursue these activities. Especially with the technology overuse and the emergence of postmodern tourist behavior, super-smart tourists (tourist 5.0) have started to become a topic of interest. Super-smart tourists are known for taking advantage of modern technology in many ways, to adapt to unexpected changes in tourism activities quickly. They utilize technology thoroughly at every step of their journey, from the very early stages of planning to the feedback after the trip. Therefore, super-smart tourists leave electronic traces and personal information on airline booking systems, hotel reservation systems, and restaurants, in addition to their traces on social media networks. The knowledge that can be derived from this provides businesses with a dynamic ability to integrate with internal and external communities (Fuchs *et al.*, 2014: 2).

The activities of tourists are recorded through various channels, and databases are created by tourism establishments. Database marketing is an approach used by tourist establishments to anticipate the need of current and potential customers to prepare products and services accordingly. Additionally, tourism establishments use websites, social media, and many other online platforms to investigate tourist behavior and their anticipation and satisfaction levels.

The collected data is further analyzed to improve the services and act when required. Tourism establishments use database marketing approaches intensively to increase efficiency and productivity, to create cross-selling opportunities, to make tourist segmentation and positioning, to predict tourist behavior, to develop products/services according to tourist expectations, and to reduce costs in their marketing activities.

To supply the smart product that the tourism industry needs, it is necessary to analyze the data produced on many devices. Activities of tourists are recorded through various channels and databases created by tourism establishments. The data produced by many devices contribute to the formation of a common communication terminology in the world (Wise & Heidari, 2019: 23). These resources, derived from database communication, reveal which hotel tourists prefer, what they eat and drink, where they want to visit, and whether they are delighted or not. The quality of data, reliability of the source, interaction, and accessibility is crucial for tourism products and services to decide on suitable strategies. The quality of the data produced at any moment increases the potential of creating big data.

Big data is defined as adding value by processing all the structured and unstructured data collected from various places, such as social media posts, blogs, photographs, videos, and log files. Big data affects all sectors, and it should not be considered separately from the tourism industry, since it contains an enormous amount of data. Indeed, big data is a critical tool in database marketing. The data collected through the internet provides convenience in many areas of tourism, such as marketing, operations, human resources, and communication. Increasing travel organizations, online experience ratings, clicks on the like button on blog sites, social media complaints, and call-center conversations require big data analytics to understand the customer's travel experience.

Big data analytics enables the emergence of meaningful and valuable information for establishments by analyzing large volumes of data. It also helps reveal consumer patterns and connections that are normally seen. Big data is used as a great tactical force in marketing to measure people's travel demands and greatly increases the sales capabilities of tourism establishments. Within the scope of big data analytics, data warehouse, data mining, data search, predictive analysis, deep learning, real-time monitoring, artificial intelligence, web crawler, internet robots, visualization, statistical process control, etc. approaches are adopted and evaluated.

This study will focus on how database marketing and big data analytics are used in the tourism industry, according to postmodern consumer anticipation. In this context, it aims to examine big data analysis and database marketing approaches in detail. The definition of database marketing and big data analysis, benefits, and challenges for tourism establishments will be evaluated. The approaches within the scope of big data analysis, which is an important strategic tool in database marketing, will be discussed, examples in the tourism industry will be revealed.

Literature review

Nowadays, the value of establishments is mostly limited by the knowledge it contains. Therefore, it is necessary to protect and increase knowledge-based resources in order to be sustainable and maintain a competitive advantage (Fuchs *et al.*, 2014: 2). The knowledge use, which has become one of the primary marketing tools, is not one-sided. Also, it is one of the most important weapons of both the producer and the consumer. Knowledge technologies are no longer used only for businesses to accomplish their strategic goals. Consumers can now create their own content and market their own ideas (Dinis *et al.*, 2019: 4). The ability of the consumer to produce data according to her own wishes has caused a change in the market structure. For this reason, it is possible to talk about a global market based on knowledge obtained from real data.

Since the raw data is not meaningful by itself, it can only be made meaningful when processed and structured. The knowledge management hierarchy points to four stages: data, information, knowledge, and wisdom. Each of these stages is more important than the previous.

First, the raw *data* is obtained and then transformed into *information* by structuring. *Wisdom* starts with planning how to use that *knowledge* subsequent to all of these processes. The stores where the data protected are called databases. In the broadest sense, a database is a repository where non-repetitive data and data related to each other are collected to provide multi-functional use (Uyar, 2016: 1691). Database marketing is an approach used by tourist establishments to anticipate current and potential customers' needs to prepare products and services accordingly. Establishments have become progressively dependent on online platforms in order to develop new products and services, gain a competitive advantage, and above all, keep up with the latest innovations. These platforms possess an important mediation role between establishments and consumers (Ferreira, 2019: 76).

Establishments can easily access knowledge about tourists' physical location thanks to technologies that they carry with them, such as smartphones, tablets, and computers. This opportunity offers various marketing advantages to establishments and stakeholders (Vassakis et al., 2019: 106). There are three crucial stages that establishments must adopt to have effective database marketing (Uyar, 2016: 1691), and these stages are:

1. **Data collection:** This stage is about collecting relevant raw data for establishments to realize their strategic activities and gain insight into the market.
2. **Transforming data into knowledge:** Effective marketing is required for a successful establishment's activities. This stage refers to converting the raw data by the establishments into knowledge from which they can obtain value.
3. **Developing and transforming business strategies:** The knowledge obtained at this stage should be used to accomplish strategic goals. As a result of the process, customer information can be divided into categories and used for strategic purposes such as market segmentation.

Additionally, tourism establishments use websites, social media, and many other online platforms to investigate tourist behavior and their anticipation and satisfaction levels. The collected data is further analyzed to improve the services and act when required. Tourism establishments use database marketing approaches intensively to increase efficiency and productivity, create cross-selling opportunities, make tourist segmentation and positioning, predict tourist behavior, develop products/services according to tourist expectations, and reduce costs in their marketing activities. When evaluating a data source's quality, its intelligibility, scope, usability, and effectiveness should be considered (Sigala *et al.*, 2019: 69).

The benefits of database marketing are as follows (Tokay, 2019: 37):

- The establishment's budget for marketing can be used more efficiently and effectively due to database marketing.
- The establishments can develop a relationship with the target customer mass.
- Customer satisfaction and loyalty can be created more easily.
- It can be produced promotions that are meaningful for current and potential customers.
- The establishments can take an opportunity to cross-sell.
- Target customer mass can be attained more easily by segmentation of the market.
- The establishments provide a competitive advantage.
- Customer demands and needs can be predicted early.
- Product development, market research, and many types of research become easier.
- It helps to regain forfeited customers.

All these benefits are directly related to the effective and efficient use of the database. Database marketing provides an essential base for customer relationship management (CRM). Database marketing, which was dependent on traditional data collection tools in the early 2000s, has now become dependent on non-ordinary sources such as big data. Since even unstructured data becomes more reliable than traditional data sources due to developing technologies, big data is crucial for database marketing. It can be said that it is an evolutionary resource in customer relationship management. Big data, which guides making conscious and dynamic decisions and predicting the future, helps extend product and service life expectancy and directly supply market needs (Stylos & Zwiegelaar, 2019: 168-169).

Big data

Data constitute the most important part of corporate life and the modern economy in business processes that are reshaped in the knowledge age we are in. In addition to data development, processing, and protection for business strategies, data is the key building block of the knowledge systems that support the establishment (Trom & Cronje, 2020: 647). The big data concept is used to describe high volumes that are difficult to store, process, and analyze with traditional database technologies (Hashem *et al.*, 2015: 100). To put it simply, it is a term that describes a high volume of structured or unstructured data (d'Amore *et al.*, 2015: 170).

Complex data in different formats with very large volumes such as photos, videos, social media content, website posts, smart sensors, customer records, GPS signals, etc. produced every second in digital environments establish the core of big data (Trom & Cronje, 2020: 647). The extensive data produced on all these platforms in a structured, semi-structured, and unstructured format expresses a continuous growth from the inside out (Hashem *et al.*, 2015: 99). Big data was initially described as 3V (volume, variety, and velocity), while veracity and value were added to them in 4V and 5V (Davenport, 2014: 7; Chen & Zhang, 2014; Centobelli & Ndou, 2019: 1). However, Vassakis *et al.* (2019: 105-106) evaluate together seven elements when defining big data:

1. **Volume:** There is the existence of datasets created by users on technologies such as the internet of things (IoT) or online platforms.
2. **Variety***:* There is a variety of sources and formats from which the data is obtained.
3. **Variability***:* There is a significant semantic change in the content of the data.
4. **Velocity***:* It is the real-time high-velocity data generation of interconnected devices.
5. **Veracity***:* Data reliability should be involved by accessing valuable data.
6. **Visualization***:* The data or the information obtained is visualized and effortlessly understood.
7. **Value***:* A value must be created with the data obtained to achieve the objectives efficiently and effectively.

When Google web searches are reviewed, searches for the phrase "Big Data" appear to be more than 290 million (Mariani *et al.*, 2018: 3515). Big data analysis is also the meaningful processing and use of this massive volume of data using database and advanced software techniques (d'Amore *et al.*, 2015: 170). To put it simply, big data analysis can be defined as the ability to convert a quantity of data into a value.

Classification of big data

Big data came into our lives very quickly with technological developments. As human beings try to understand what big data is, it has become more complex thanks to increased data roaming (Trom & Cronje, 2020: 645). The classification of big data obtained in different formats in various sources is seen in Figure 9.1.

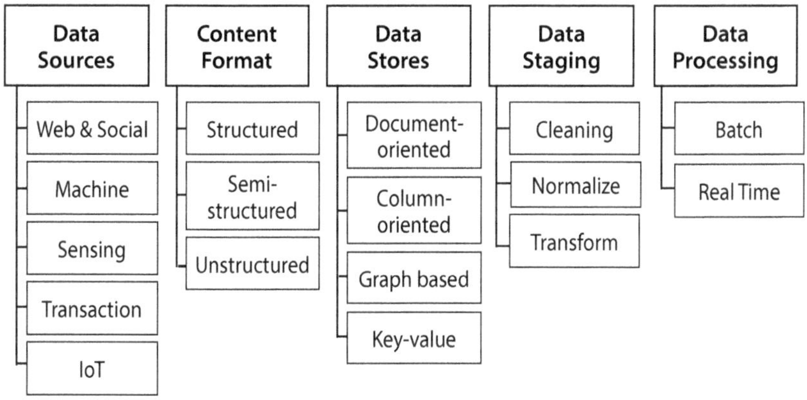

Figure 9.1: Big data classification. *Source*: Hashem *et al.* (2015: 101).

Big data is classified according to five categories: data sources, content format, data stores, data staging, and data processing (Hashem *et al.*, 2015: 102; Fuchs *et al.*, 2014; Shereni & Chambwe, 2019: 2):

Data sources:

- *Web and social:* Data generated through information and ideas shared by users in virtual communities created in areas such as websites and social media (Blog, Facebook, Twitter, etc.).
- *Machine:* Data produced by hardware and software on machines and devices without human factor.
- *Sensing:* Data produced by devices developed to measure physical features and detect them through signals.
- *Transaction:* The data that the transaction reveals. It is a set of events that contain financial, transactional, and business data, and describe them in a time dimension.
- *IoT:* An enormous amount of data that occurs because of communication between each other of internet-connected devices used to fulfill various needs.

Content format:

- *Structured:* Data that is consciously collected and stored to manage, query, and use as information through a conventional database.For example of transaction data, survey, and rating data.
- *Semi-structured:* Data that cannot be tracked with a conventional database and is in an unarranged format.
- *Unstructured:* Unstructured data obtained through websites and social media that does not have a specific format. The type of data is more difficult to analyze and make sense of than data from other sources.

Data stores:

- *Document-oriented:* These resources use assorted forms (XML, JSON, etc.) as data stores to store documents and information.
- *Column-oriented:* Databases that store data in columns, based on their characteristics.
- *Graph based:* Systems where data is stored by identifying the correlation between models through chart models.
- *Key*-value: It is a system used to deal with an immense database. In this system, data is stored and associated with the key value.

Data staging:

- *Cleaning:* A process of identifying missing or meaningless fields and rows.
- *Transforming:* The process of making data suitable for analysis.
- *Normalization:* The process of minimizing data redundancy in the data set and configuring the database.

Data processing:

- *Batch:* Refers to systems that allow clustered and linked databases to be made into scalable clusters.
- *Real Time:* Systems that support, store, and filter unlimited real-time data flows.

The data obtained can cover a number of these classifications. Therefore, establishments must master this classification so that their databases can be used effectively. Thereby, an establishment that takes data from any source in any format and at any intervals, and stores it and analyzes it thoroughly, will achieve success in a short time.

Benefits and challenges of big data

With the development of data science and the emergence of big data, the way establishments and government administrations observe and analyze the market has altered rapidly (Vassakis *et al.*, 2019: 105). Big data has revolutionized establishments' business models and performance metrics (Sigala *et al.*, 2019: 57). The main reason for this is that databases, which are a substantial resources, flow towards the establishments not only from the establishments themselves, but also from many various sources. Nowadays, the world's most accomplished companies are the first to realize the potential to turn enormous amounts of data from a variety of sources into marketable value. Tech companies like IBM, Google, and Facebook were among the first to understand big data's potential value (Trom & Cronje, 2020: 647). These companies have tracked user information and users' footsteps in the digital environment, using this data to anticipate and guide their expectations and desires. Thanks to advanced big data analysis, they have been among the fastest developing establishments, and with a high profit margin.

However, while big data was only of interest to multinational companies in the early days, it began to attract the attention of medium and small-scale companies with the awareness of the importance of knowledge today (Trom & Cronje, 2020: 645), because big data offers essential opportunities to provide high-quality service and work with efficiency (Song & Liu, 2017: 15-16). Despite being a good tool to support decision-making, it is still in its start-up phase (Stylos & Zwiegelaar, 2019: 168). Today's establishments are now wondering not only about current developments and the reasons for these developments but also what will happen in the future and what situations will emerge (Trom & Cronje, 2020: 645-646). Thanks to big data analysis, organizational agility and competitive advantage can be achieved by improving information management (Espinet, 2019: 33).

Big data provides excellent insight into customers' recognition and analysis (Song & Liu, 2017: 15). It has also become a crucial resource for activities that require business intelligence, such as finding new consumers, representing consumers, and gaining loyalty (Mariani *et al.*, 2018: 3515), because big data is not limited to a single establishment or region. Thanks to big data targeting the masses, achieving social or economic development goals and measuring these activities are possible. It is also possible to complete existing gaps or shortcomings in the official data of states, local governments, and establishments (Wise and Heidari, 2019: 24).

Developed technologies such as machine learning (ML), Internet of Things (IoT), and artificial intelligence (AI) are the primary resources that support big data. These intelligent technologies help the service process owing to the information they receive from it while feeding big data. The technologies improve the marketing process with their ability to solve problems and ambiguities (Inanc-Demir & Kozak, 2019: 220). However, companies that adopt the big data business model act with different purposes, and are mainly focused on gaining a competitive advantage by reducing costs (Espinet, 2019: 33).

It is difficult to process and analyze with traditional statistical and software methods given the characteristics of the data collected on these various platforms (Mariani *et al.*, 2018: 3515). Despite this difficulty, this enormous data must be converted into knowledge (Jackson, 2014: 15). Otherwise, establishments will get nothing but enormous data dumps. Hence, establishments need to have modern management and analysis competence to turn big data into a competitive tool by making it meaningful. However, many businesses that follow developments are aware of big data but do not know how to manage it and analyze it (Sigala *et al.*, 2019: 58). However, establishments are not inclined towards big data strategies because of the high costs that data analysts initially incur (Shereni & Chambwe, 2019: 2).

Big data is a process of discovery that needs to be studied attentively. Otherwise, the findings may be meaningless or deceptive. So, the intensive discovery process has a high risk (d'Amore *et al.*, 2015: 171). Establishments now have high volume and velocity of data from a variety of sources. In this case, the conventional data processing systems we are accustomed to cannot deal with the present's data (Trom & Cronje, 2020: 645). Despite all these developments, there are a set of factors that preventing the utilization of big data (Sigala *et al.*, 2019: 62):

- Limited ability to interpret and manipulate big data, including the capacity and similarity of data sets, data sources, and technological platforms to be used.
- Limited time allocated to personnel for big data analysis.
- Budget constraints for performance measurement.
- Concerns about ethics, security, and reliability in the use of big data.
- Procedures of governments and institutions restricting access to data.

It also includes a troublesome process to analyze due to the properties of big data and these elements that hinder its utilization. Data mining engineers may have difficulty in dealing with immense quantities of data.

In addition, the lack of data scientists in the field also challenges big data marketing (Shereni & Chambwe, 2019: 5-6). The great concerns for utilizing big data analytics in marketing are privacy and security (Espinet, 2019: 33). In short, as well as the advantages it provides, it has many risks such as privacy, ethics, and reliability (Sigala *et al.*, 2019: 58). Insufficient information, inability to process, and misleading data are also drawbacks (Trom & Cronje, 2020: 646).

The challenges of big data, according to the lifecycle, are classified as data, process, and management by Sivarajah *et al.* (2017: 265):

1. **Data challenges**: Challenges arising from the properties of data. The volume, velocity, variety, variability, veracity, visualization, and value of the data identify these challenges. To deal with these, the source of the data must be well known, and must be reliable.
2. **Process challenges**: These are the technical challenges of data. Challenges to acquisition and warehousing, process data mining and cleaning, aggregation and integration, analysis and modeling, and data interpretation. For the big data process to be well managed, it is essential to choose and execute a well-crafted business model.
3. **Management challenges:** These draw attention to issues such as data ownership, cost/operational expenditures, data sharing, and data governance during data management, necessitating good managerial skill.

The challenges during big data directly affect data analysis and process. However, the strategies developed by establishments to deal with these challenges will result in great costs in the short term, but they will be able to bear these costs in the long term.

Big data analytics and process

Big data analysis gives establishments the opportunity to explore their own potential. It is important for them to rediscover what their establishment's potential is, how they can use it, what strategies they need to develop (Sigala *et al.*, 2019: 58). Big data does not offer great solutions and successes in a short time, as expected. Establishments that invest ought to make short and long term plans to assess whether big data is essential and then take action. The failure of the business to integrate database operations and the lack of management practices can lead to the failure of big data investments. Businesses first need to gain experience with big data and experience the trial process in it (Espinet, 2019: 33). In order to generate value with big data, the following strategic business process must be adopted (Sigala et al., 2019: 65-66):

- **Step 1:** It is the step in which big data initiatives and investments start. In this step, a strategic policy should be adopted to address this problem by identifying a question first, and the benefits of big data should be determined. First of all, the problem really needs to be determined correctly. Especially when the big data strategy is executed with outsourcing, the feedback becomes meaningless when the problem is not determined correctly, or the desired thing is not fully transferred, and the big data attempt fails. Several questions should be improved for the problem situation, and the points that big data is expected to solve should be adequately determined. For this, a project team for big data should be established. Variables for problems, such as establishing a research hypothesis, should be defined, and the indicators that can be measured for each variable should be explained, and data sources for these indicators should be identified.
- *Step 2:* At this step, initiatives for big data need to be implemented. Data must be taken in data sources and classified according to specific properties. In particular, data ought to collect in accordance with the variables to be analyzed. Analytical and statistical methods should be applied, and the findings obtained filtered. Besides, the benefits, costs, and reasons of the analysis methods to be used at this stage should be determined correctly, and the superior and missing points of the alternatives of the decision analysis method addressed. In conclusion, the information obtained from the data ought to visualize.
- *Step 3:* Knowledge from big data at this stage ought to be put into action. Action should be taken by adopting an informed decision-making approach, and tangible/intangible results should be managed. The information obtained in the second step ought to be used for change and development. Otherwise, this whole process will bring nothing but cost to the establishment.

Similar to these three steps, big data analysis includes four processes: data sources, data extraction, data warehousing, and data mining, according to Fuchs *et al.* (2014: 3). Establishments should make data-driven decisions when they need to make a substantial decision. For this purpose, they have to use various analysis methods to make meaningful the data they obtain from various sources. While it has infinite potential in big data, the analysis of this data can be done with restricted technologies and tools (Sivarajah *et al.*, 2017: 265). Modern establishments that adopt smart technologies have adopted data mining to get value with big data (Shereni & Chambwe, 2019: 2). Data mining is the process of obtaining understandable information from an enormous amount of data. This process is also called knowledge discovery in the database (Rong, 2014: 3659). This discovery can be seen as a treasure hunt because the cornerstone of the era we are in is knowledge.

Data mining technologies include Google analytics, signal processing, genetic algorithms, cluster analysis, time series analysis, data fusion and integration, natural language processing, machine learning, network analysis, data mining, spatial analysis, simulation, and visualization (Shereni & Chambwe, 2019: 6). One of the most common big data analyses is social media analytics. This kind of analysis is carried out through social media channels with structured or unstructured data. Platforms such as Facebook, LinkedIn, Twitter, Instagram, Flickr, and YouTube are the most discussed in these analytics. In these applications, users' comments about a situation or the communities they follow are analyzed (Vassakis *et al.*, 2019: 106-107).

Data analytics primarily consists of four stages: data collection, data extraction, data warehousing, and knowledge generation (Sigala *et al.*, 2019). The two crucial processes are data management and data analysis. During the data management phase, it is well determined which data will be collected in which ways, how to clean up irrelevant data, and how to integrate data into the system to generate value. The data analysis phase means the determination and interpretation of accurate data modeling and analysis techniques (Sigala *et al.*, 2019: 65).

Big data and tourism

The tourism industry has a significant share of the world's overall exports and market share, in addition to being a labor-intensive economic activity. Therefore, tourist destinations, suppliers, demand forecasts, tourist trends, and all other factors have to be planned and programmed in a long-term way (Höpken *et al.*, 2017: 188). With the development of the hospitality and tourism industry, the importance of scientific decision-making is increasing to gain a competitive advantage in the market. Increased data sources with high volume, velocity, and variety have affected all industries as well as tourism. Decision-makers need new tools and capabilities to capitalize on these enormous resources (Ferreira, 2019: 75). Hence, it has become inevitable to discover big data in the hospitality and tourism industry with high potential methods such as data mining (Rong, 2014: 3659).

The new generation, which we call the postmodern tourist or tourist 5.0, maintains uncertainty at the highest level and moves with completely complex emotions. This consumer, who focuses on symbol and image, and is both a producer and a consumer, is active as well as displaying a contrasting opinion. This unique generation, which wants to be liberated by moving away from social constraints contrary to meta, status, and norms, acts inconsistently, ambivalently, individually, and irrationally (Odabaşı, 2017: 125). These behaviors of consumers are reflected in tourism activities. In this case, more robust forecasting methods and technologies are needed to deal with such an undetermined tourist profile.

Together in technological developments, information production and consumption experienced a transformation in another form. Tourists can create their own travel content with the information they collect from various sources and, at the same time, share their opinions and experiences on various platforms. Before all these developments, tourism enterprises and stakeholders were the only sources of knowledge. That is why tourists were making their travel plans, relying on the information they shared. Now content for tourist experiences on social media has become more reliable than the information offered by establishments (Alaei *et al.*, 2017: 1). In short, it can be said that a large part of tourism activities have transferred to the digital environment. From choosing the travel route to booking rooms and tickets, from check-in to sharing experiences, many processes take place in the digital environment (Dinis *et al.*, 2019: 2). All transactions and activities made by tourists based on the internet leave a digital footprint (Alaei *et al.*, 2017: 1). Tourists are seen to use multiple web pages as a source of information when making decisions about their travels (Dinis *et al.*, 2019: 4). Tourists aspire to assure themselves by obtaining the information they need from various sources. However, in return for this request, they leave their footprints on their web pages, even without realizing it and offer data miners much information about themselves. That is why an enormous amount of data is being produced for the tourism industry at any moment (Vassakis *et al.*, 2019: 107).

Big data contributes significantly to the accurate forecasting of tourism demand. Providing the necessary information to destination management, touristic product suppliers, and residents are essential for determining the right strategies and regulating price and supply policy. As an economic determinant in traditional tourist arrival forecasts, consumer price indices, prices of regional products, price policies implemented by competing regions, interest and unemployment rates are used, while advertising activities, crisis situations, natural disasters and similar situations were considered. Today, however, tourist demand forecasts are made using web-based data sources such as search engine traffic and customer feedback online. These resources are characterized as big data and lead to a significant improvement in forecast performance (Höpken *et al.*, 2017: 189-190). Therefore, it can be said that big data offers rational and critical perspectives in describing the current tourism phenomenon and exploring its dynamics (d'Amore *et al.*, 2015: 179).

Online platforms are convenient for promoting tourism activities (d'Amore *et al.*, 2015: 169-170). The hospitality and tourism industry focuses on creating value for people and keeping that value alive through experiences. Tourist experiences begin on these platforms. Thanks to the data produced on these platforms, it is easier to understand what is important, and value is created with the data obtained from people (Trom & Cronje, 2020: 647). Tracking tourist behavior in tourism marketing and the correct working of a feedback mechanism depends on the information's accuracy (Song & Liu, 2017: 20), because activities, such as ensuring tourist satisfaction, tourist loyalty, and sales development within the scope of customer relations management in tourism establishments, are carried out based on this information (Talón-Ballestero *et al.*, 2018: 187). Dynamic packaging and recommender systems are supported by big data, and personalized service is offered to tourists (Höpken & Fuchs, 2016: 1). New business strategies, new products/services, and marketing campaigns can be developed with data providing insight into travel activities (Alaei *et al.*, 2017: 1).

If big data is managed effectively and strategically by tourism establishments, it increases enterprises' competitive advantage and enables them to succeed (Trom & Cronje, 2020: 647). Due to its ability to fulfill modern marketing requirements, it can untie the crucial problems of establishments (d'Amore *et al.*, 2015: 170). To enjoy the benefits of digital platforms, the marketing departments of tourism establishments must focus heavily on the online environment that delivers this huge data source (Khmiadashvili, 2019: 923). A tourism establishment collects and processes big data with universal access to achieve specific goals. However, this opportunity also requires considered expenditure and competence. While large-scale establishments can provide the budget to meet these requirements, this is not the case for small establishments (d'Amore *et al.*, 2015: 179). However, big data is more advantageous than traditional data collection and analysis methods.

These advantages for tourism establishments using big data are as follows (Song & Liu, 2017: 16-17):

1. **Reliability***:* More reliable results can be achieved due to the analysis of tourists' actions. Thanks to big data, tourist behavior is analyzed using multiple sources instead of single sources, and the market can be shaped.
2. **New knowledge flow**: The big data that emerges in tourism is the rich data produced by tourists. There is a constant flow of new information from big data to meet ever-changing tourist expectations and demands.
3. **Real-time data**: Big data mediates in obtaining real-time data of tourists. Tourist behavior can be identified as the moment it occurs and responded to instantly.

The optimum use of big data in both operation and destination management allows the right service to be provided to meet the expected demand and contributes to the smart tourist destination (Dinis *et al.*, 2019: 3). It points to a smart destination based on smart tourism activities. Smart destinations attract tourists' attention with their high competitiveness, openness to innovation, and high living standard. Data-driven platform services for infrastructure, economy, society, and the environment have become the most substantial aspects of attractions (Wise & Heidari, 2019: 22). In addition, it is not enough to use a single data source to examine tourist activities in the destination, but the data obtained from several different sources are complementary to each other (Salas-Olmedo *et al.*, 2018: 13).

Which destinations tourists visit during their trips is very important information for destination marketing organizations. Marketers can more easily unlock destination potential through this information from various sources. Destinations can generate routes and make the region more attractive by doing regional co-marketing activities (Önder, 2017: 55). Smartphones and tracking technologies that support big data are the most powerful tools for evaluating to provide personalized service by comprehending tourists' behavior and travel motives (Rodríguez *et al.*, 2018: 1; Espinet, 2019: 31). If tourist destination and establishments want to increase or maintain competitiveness, they have to adapt to technological changes and keep track of continuously generated data (Wise & Heidari, 2019: 23).

The tourism marketing mix and big data

Tourism provides many sources and application models in big data collection (Guan & Du, 2016: 353). Internet browsers and smart mobile devices have become the primary source of data for establishment owners, managers, local governments, residents, and tourists (Wise & Heidari, 2019: 23). Big data collected from various sources is used to fix urgent problems and reveal potential value at an affordable cost (Irudeen & Samaraweera, 2013: 207). However, while developing AI systems that feed big data, consumer behavior models and marketing strategies have changed structurally (Inanc-Demir & Kozak, 2019: 213). With these developments, big data triggered the tourism marketing mix within another and caused its reshaping (Stylos & Zwiegelaar, 2019: 168). The traditional tourism marketing mix (4P) is expected to be shaped like this with big data (Inanc-Demir & Kozak, 2019: 217-220):

1. **Product***:* Thanks to systems such as IoT, AI, and ML that support big data, tourists can be contacted more directly. Establishments can get ideas about tourist product development, presentation, and feedback, while at the same time identifying tourist profiles and behaviors, tourist flow, routes, and tourist safety. All this data has a significant impact throughout the process of ingesting the development and delivery of the personalized product to the tourist.
2. **Promotion***:* After the development of touristic products and services, target audiences should be focused on. Personalized promotional activities can be provided through AI applications by virtue of big data. The promotional activity process can be initiated according to tourists' consumption habits and trends, hobbies, education levels, gender, age, behaviors, and destinations they have visited before. Moreover, this promotional process is carried out effortlessly owing to artificial intelligence applications.
3. **Price***:* Big data has also had an impact on the pricing strategies of establishments. With price optimization analytics, the establishment's profitability and costs can be calculated automatically and the value to be added to the product determined. It also contributes to the release of market values of big data products and services. Europe's major airlines, such as Air France and KLM, for example, use automated pricing systems that addressed the origin and destination points of journeys over two years of data and price by estimating possible cancellation, change, and double-booking rates.

4. **Place:** the concept of place has transformed due to the intensive utilization of information in the service industry. Now place in marketing means digital environment. The best example of this is to make hotel/plane reservations and check-ins digitally, and due to digital ID cards, all services can be provided to hotel guests in a personalized way by looking at their past experiences.

Big data analytics and tourism

The hospitality and tourism industry is intertwined with other industries and has a susceptible structure. Developments or tensions in other areas directly affect the tourism industry and lead to a change in tourist behaviors and trends. Therefore, high-tech methods such as AI have been used to predict tourism demand, which has a very flexible structure (Höpken *et al.*, 2017: 190). Survey practices, the traditional method for measuring and evaluating tourist behavior, are inadequate to reflect reality and have a high rate of bias (Rodríguez *et al.*, 2018: 2). Unlike the survey method, big data sources that provide data from users' real actions also increase the accuracy of tourism market research (Song & Liu, 2017: 16). The seasonality of tourism and its great impact on demand fluctuations also significantly affect its measurement. However, through advanced technologies, it is now possible to verify tourist numbers and characteristics (Yamamoto, 2019: 157). Höpken *et al.*'s (2017) found that big data analysis gives better results, and predicting performance is better by taking advantage of data mining. Talón-Ballestero *et al.* (2018:189) emphasize the importance of big data in customer profile analytics.

Tourist behavior and preferences can be exposed through big data, data mining, machine learning, artificial intelligence, and semantic web and social networks, which are especially substantial for the field of tourism marketing (Yamamoto, 2019: 158). Big data sources, structured or unstructured, are used to focus on changing tourist preferences in destination management (Vassakis *et al.*, 2019: 106). Web browsing and GPS data can be captured and embedded in a tourist destination. This massive data, which has a very high potential to turn into valuable knowledge, is often not evaluated. The foremost reason for this is that destination management does not have an information infrastructure to evaluate this actual data (Fuchs *et al.*, 2014: 1).

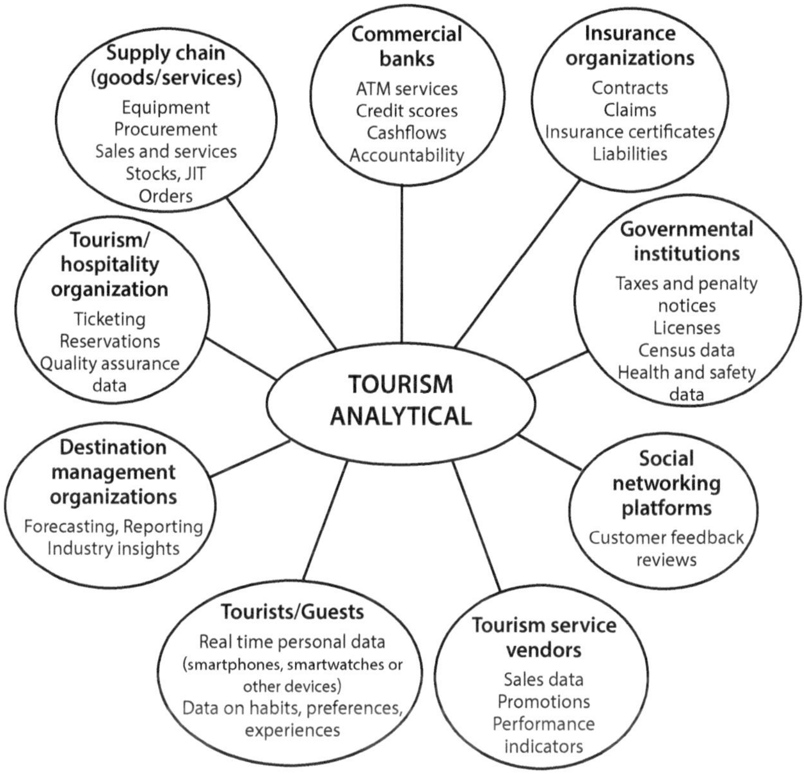

Figure 9.2: Big data tourism analytical. *Source*: Stylos and Zwiegelaar (2019:175).

Data in various formats from commercial banks, insurance organizations, governmental institutions, social networking platforms, tourism services vendors, tourist, destination management organizations, tourism/hospitality organizations and the supply chain provides a rich flow of data about tourists and consumers. This data flow is the first step for big data analysis in tourism. This supports big data infrastructure through experts such as software developers, web designers, database architects, data consultants, and business intelligence managers to identify tourist profiles and provide personalized services. Big data-driven decision-making mechanisms will be supported in the tourism ecosystem due to the necessary business intelligence and improved analytical capabilities of these experts. In this way, the data storage required by the hospitality and tourism industry and the big data platform in the tourism industry will supply the necessary procedures in order to carry out an effective analysis. The big data service supports tourism through analytics systems such as blockchain, cloud systems, etc., to store and analyze data. Customer relationship management is supported by the established big data infrastructure and database management system (Stylos & Zwiegelaar, 2019: 175-176). Real-time and accurate data flow, such as ATM Services, which initially seemed very small, will reshape both businesses and the hospitality and tourism industry's performance.

In order to talk about smart tourism, it is important to make both operations and products smart by the new business models. Products, services, and operations must be harmonized using, storing, processing, and analyzing data (Dinis *et al.*, 2019: 3). The important thing is that businesses and destination management organizations evaluate their feedback correctly and determine their strategies accordingly. Fuchs *et al.* (2014: 3) state that four main steps: data sources, data extraction, data warehousing, and data mining, must be carried out in order for a destination that supports the specified model to qualify as an information destination. Text summary and sensitivity analyses, along with data mining, are among the valuable approaches to determining tourist behavior in the hospitality and tourism industry (Alaei *et al.*, 2017: 3).

Applicability of big data in tourism

The smart city has become a popular trend, and the smart tourism activities associated with it are the key to urban sustainability. Sustainable principles should be based on minimizing the adverse effects of tourism activities in social, economic, psychological, and environmental directions and to ensure prosperity in all directions. In order to utilize less energy, ensure ecological integrity, and talk about social justice, the possibilities offered by technology ought to be fully exploited. Just as the quality-of-service operations decrease when a hotel or a bus exceeds its carrying capacity, the quality of life in the destination decreases when a destination exceeds its carrying capacity. With these high technologies, destination carrying capacity can be controlled by the destination data flow. For this, big data infrastructure ought to develop by connecting urban applications, smart buildings, light and traffic sensors, smart boards, meters, and many more data sources in the destination (Wise & Heidari, 2019: 23).

One of the important factors in the adoption and availability of big data is open and closed approaches in different countries (Pesce *et al.*, 2019: 19). Spain, for instance, adopts an open approach, integrating big data into all tourism systems. In particular, online travel agencies or tourism establishments need steady infrastructure to make the data they collect through their websites sustainable. Otherwise, if there is no legal regulation and information security is not ensured, existing resources will be used in vain. For the effective use of resources, it must be worked by the staff with the necessary professional expertise and experience (Espinet, 2019: 53).

The utilization of state-of-the-art technologies by tourism destinations and establishments will also facilitate big data analysis. Technologies such as IoT, AI, and ML are extremely successful at collecting and controlling data. Machines create an open interconnected communication environment via IoT systems (Inanc-Demir & Kozak, 2019: 215). In the marketing of touristic products and services, IoT supports the business and destination to obtain accurate and quality information by increasing data production. At the same time, these systems allow simultaneous responses to need and demands (Wise & Heidari, 2019: 24). AI is another state-of-the-art technology that supports big data. Sentiment analysis can be done, and consumer behavior can be controlled via AI applications. It is a driving force for the collection, storage, and utilization of data when required.Through sentiment analysis, consumers' views on tourism and touristic products & services are explored. These analyzes are seen as an opportunity to examine the purchasing decisions of potential consumers for touristic products & services.

ML, on the other hand, is an AI sub-segment that computers learn from. Technology can produce and use its data itself due to ML. Technology companies like Facebook and Google are widely using ML to take market advantage. In addition, these technologies have turned upside down the way they do business in finance, health, safety, and many other sectors (Inanc-Demir & Kozak, 2019: 216).

The tourism industry is faced with unimaginable degrees of knowledge. In particular, platforms such as TripAdvisor, Booking, Expedia, Trivago, and Airbnb voluntarily collect data with user reviews. Users who follow these platforms decide their travels considering the shared experiences of tourists. These platforms, which have been an important road to tourists, have been influential in tourist product marketing (Ferreira, 2019: 76). Since these sites contain a tremendous amount of data, they are a knowledge mine for the tourism industry. Generally, tourism research use data such as geo-tags, photos, internet history, and GPS. These sources provide the opportunity to follow consumers in depth. In addition, these resources supply a rich source of data to understand tourist attractions, timeframe, and behavior (Salas-Olmedo *et al.*, 2018: 14). Free Wi-Fi service, available especially at airports, destinations, and hotels, is an excellent opportunity to collect and analyze data.

Amadeus is one of the companies that actively uses big data and is aware of its advantages. Global distribution systems such as Amadeus contain the data of passengers traveling all over the world, the airlines used, car rentals and hotel bookings. Amadeus has an immense amount of passenger data within its company and uses it to develop its products and services, reduce costs, respond to customers' requests in less time (the answer to a request can be given in 300 milliseconds), and determine their policies (Davenport, 2014: 84-85).

When the literature is examined, there is a lot of research on the applicability of big data in the tourism industry. For instance, researchers using the text analytics approach focused on hotels and conducted big data analytics with consumer reviews from platforms such as Expedia.com. Japan Tourism Agency used the international roaming service in 2014 to access information such as location and time (Yamamoto, 2019: 158). Salas-Olmedo *et al.* (2018) compared the data received via Panoramio, Foursquare, and Twitter to examine tourist activities and determine tourist densities according to digital footprints (Önder, 2017). Marine-Roig and Anton Claye (2015) highlights the benefits of big data analysis to support smart destinations, in their research. Luo *et al.* (2018) investigated tourist behavior in gaming destinations according to social media check-ins conducted at a venue. In addition, when research in the field of tourism is examined, data can be obtained for the consumer through enterprises that provide panel data, such as Amazon. All these studies reveal the applicability and benefits of big data in the tourism industry.

Conclusion

The characteristics of the hospitality tourism industry bring many restrictions. Tourism activities are significantly affected by geographical location and seasonal structure. Data showing a high rate of spatial and time distribution are needed to study and diversify tourism activity. It can be said that big data is an appropriate tool for filling this gap (Silva *et al.*, 2018: 1020). To cope with these challenges, developed countries such as China, Italy, and the United States contribute significantly to big data research (Centobelli & Ndou, 2019: 7). Tourism is an experience economy. Tourist experiences help businesses create value. In order to improve these experiences, it is necessary to see the implicit aspects of the pleasure and emotions of tourists. For instance, observing tourists' behavior in the hotel, where they spend a long time, offers an important tip to the managers. Also, big data is an important data source for tracking new search-powered trends such as Google, Yahoo, and Bing, and for creating new trends.

To do database marketing practices in the hospitality and tourism industry, both establishments and destinations need to build their technology infrastructure. Technologies that act as virtualized stores, such as cloud computing will support these investments of establishments. Through IBM, Microsoft Azure, and Amazon AWS, data storage, protection, and the service offered can be integrated at less cost (Hashem *et al.*, 2015: 99). In addition, it is imperative that tourism establishments take big data potential seriously, form a team and work on it meticulously. Otherwise, labor and time will be wasted, resulting in severe costs. It should not be forgotten that to create a smart tourist destination, all stakeholders must first take smart steps, and big data is the first of these smart steps.

Discussion questions

1. What are the barriers to database marketing and big data applications in the hospitality and tourism industry?
2. Could database marketing and big data create a new field of work in the hospitality and tourism industry?
3. What are the tourist applications that feed big data?
4. How can hospitality and tourism enterprises deal with big data?

References

Alaei, A.R., Becken, S. & Stantic, B. (2017) Sentiment analysis in tourism: Capitalizing on big data, *Journal of Travel Research*, 1-17.

Centobelli, P. & Ndou, V. (2019) Managing customer knowledge through the use of big data analytics in tourism research, *Current Issues in Tourism*, 1-21.

Chen, C.L.P. & Zhang, C.Y. (2014) Data-intensive applications, challenges, techniques and technologies: A survey on big data, *Information Sciences*, **275**, 314-347.

d'Amore, M., Baggio, R. & Valdani, E. (2015) A practical approach to big data in tourism: A low cost raspberry pi cluster, in I. Tussyadiah, and A. Inversini (ed.), *Information and Communication Technologies in Tourism*, Switzerland: Springer International, pp. 169-181.

Davenport, T.H. (2014) *Big Data at Work: Dispelling the myths, uncovering the opportunities*. Boston, Massachusetts: Harvard Business Review Press.

Dinis, M.G.F., da Costa, C.M.M. & Pacheco, O.M.D.R. (2019) Composite indicators for measuring the online search interest by a tourist destination, in M. Sigala, R. Rahimi, & M. Thelwall (ed.), *Big Data and Innovation in Tourism, Travel, and Hospitality: Managerial approaches, techniques, and applications*, Singapore: Springer Nature Ltd., pp. 1-20.

Espinet, J.M. (2019) Big data in online travel agencies and its application through electronic devices, in M. Sigala, R. Rahimi, & M. Thelwall (ed.), *Big Data and Innovation in Tourism, Travel, and Hospitality*, Singapore: Springer Nature, pp. 31-55.

Ferreira, D. (2019) Research on big data, VGI, and the tourism and hospitality sector: concepts, methods, and geographies, in M. Sigala, R. Rahimi, & M. Thelwall (ed.), *Big Data and Innovation in Tourism, Travel, and Hospitality*, Singapore: Springer Nature, pp. 75-85.

Fuchs, M., Höpken, W. & Lexhagen, M. (2014) Big data analytics for knowledge generation in tourism destinations – A case from Sweden, *Journal of Destination Marketing & Management*, 1-12.

Guan, D. and Du, J. (2016) Cross-media big data tourism perception research based on multi-agent, in Y. J. (ed.), *Proceedings of the 2015 Chinese Intelligent Systems Conference Lecture Notes in Electrical Engineering*, Berlin Heidelberg: Springer-Verlag, pp. 353-360.

Hashem, I.M., Yaqoob, I., Anuar, N.B., Mokhtar, S., Gani, A. & Khan, S.U. (2015) The rise of big data on cloud computing: Review and open research issues, *Information Systems*, **47**, 98-115.

Höpken, W. & Fuchs, M. (2016) Introduction: Special issue on business intelligence and big data in the travel and tourism domain, *Information Technology & Tourism*, **16**, 1-4.

Höpken, W., Ernesti, D., Fuchs, M., Kronenberg, K. & Lexhagen, M. (2017) Big data as input for predicting tourist arrivals, in R. Schegg, & B. Stangl (ed.), *Information and Communication Technologies in Tourism*, Springer International Publishing AG, pp. 187-199.

Inanc-Demir, M. & Kozak, M. (2019) Big data and its supporting elements: Implications for tourism and hospitality marketing, in M. Sigala, R. Rahimi, & M. Thelwall (ed.), *Big Data and Innovation in Tourism, Travel, and Hospitality*, Singapore: Springer Nature, pp. 213-223.

Irudeen, R. & Samaraweera, S. (2013) Big data solution for Sri Lankan Development: A case study from travel and tourism, *International Conference on Advances in ICT for Emerging Regions* (ICTer), pp. 207-216.

Jackson, S. (2014) Prediction, explanation and big(ger) data: A middle way to measuring and modelling the perceived success of a volunteer tourism sustainability campaign based on nudging, *Current Issues in Tourism*, 1-16.

Khmiadashvili, L. (2019) Digital marketing strategy based on hotel industry study in Tbilisi, *International E-Journal of Advances in Social Sciences*, **V** (14), 922-927.

Luo, J. M., Vu, H. Q., Li, G. & Law, R. (2018) Tourist behavior analysis in gaming destinations based on venue check-in data, *Journal of Travel & Tourism Marketing*, 1-12.

Mariani, M., Baggio, R., Fuchs, M. & Höepken, W. (2018) Business intelligence and big data in hospitality and tourism: A systematic literature review, *International Journal of Contemporary Hospitality Management*, **30** (12), 3514-3554.

Marine-Roig, E. & Anton Clavé, S. (2015) Tourism analytics with massive user-generated content: A case study of Barcelona, *Journal of Destination Marketing & Management.*

Odabaşı, Y. (2017) *Postmodern Pazarlama: Tüketim ve Tüketici*, 5th Baskı, İstanbul: Kapital Medya Hizmetleri A.Ş.

Önder, İ. (2017) Classifying multi-destination trips in Austria with big data, *Tourism Management Perspectives*, **21**, 54-58.

Pesce, D., Neirotti, P. & Paolucci, E. (2019) When culture meets digital platforms: Value creation and stakeholders' alignment in big data use, *Current Issues in Tourism*, **22** (15), 1883-1903.

Rodríguez, J., Semanjski, I., Gautama, S., Weghe, N. V. & Ochoa, D. (2018) Unsupervised hierarchical clustering approach for tourism market segmentation based on crowdsourced mobile phone data, *Sensors*, **18** (2972), 1-17.

Rong, Y. (2014) Research of the tourism marketing basing on the data mining, *Applied Mechanics and Materials*, 3659-3662.

Salas-Olmedo, M. H., Moya-Gomez, B. & García-Palomares, J. C. (2018) Tourists' digital footprint in cities: Comparing big data sources, *Tourism Management*, **66**, 13-25.

Shereni, N. C. & Chambwe, M. (2019) Hospitality big data analytics in developing countries, *Journal of Quality Assurance in Hospitality & Tourism*, 1-9.

Sigala, M., Beer, A., Hodgson, L. & O'Connor, A. (2019) Big data for measuring the impact of tourism economic development programmes: A process and quality criteria framework for using big data, in M. Sigala, R. Rahimi, & M. Thelwall (ed.), *Big Data and Innovation in Tourism, Travel, and Hospitality*, Singapore: Springer Nature, pp. 57-73.

Silva, F. B., Marin Herrera, M. A., Rosina, K., Barranco, R. R., Freire, S. & Schiavina, M. (2018) Analyzing spatiotemporal patterns of tourism in Europe at high-resolution with conventional and big data sources, *Tourism Management*, **68**, 101-115.

Sivarajah, U., Kamal, M. M., Irani, Z. & Weerakkody, V. (2017) Critical analysis of big data challenges and analytical methods, *Journal of Business Research*, **70**, 263-286.

Song, H. & Liu, H. (2017) Predicting tourist demand using big data, in D.F.Z. Xiang (ed.), *Analytics in Smart Tourism Design, Tourism on the Verge*, Switzerland: Springer International, pp. 13-29.

Stylos, N. & Zwiegelaar, J. (2019) Big data as a game changer: How does it shape business intelligence within a tourism and hospitality industry context? in M. Sigala, R. Rahimi, & M. Thelwall (ed.), *Big Data and Innovation in Tourism, Travel, and Hospitality*, Singapore: Springer Nature, pp. 163-181.

Talón-Ballestero, P., González-Serrano, L., Soguero-Ruiz, C., Muñoz-Romero, S. & Rojo-Álvarez, J. R. (2018) Using big data from customer relationship management information systems to determine the client profile in the hotel sector, *Tourism Management*, **68**, 187-197.

Tokay, G. (2019) *Müşteri İlişkileri Yönetiminde Veri Tabanlı Pazarlamaya Dair Bir Araştırma*, master's thesis, Istanbul: Bahçeşehir University.

Trom, L. and Cronje, J. (2020) "Analysis of Data Governance Implications on Big Data", in K Arai and R Bhatia (ed.), *Advances in Information and Communication*, Switzerland: Springer Nature Switzerland AG., pp. 645-654.

Uyar, A. (2016) Veri tabanlı pazarlamanın işletmelere sağladığı yararlar: Veri tabanı kullanan kobiler üzerine bir araştırma, *International Journal of Social Research*, **9** (42), 1690-1698.

Vassakis, K., Petrakis, E., Kopanakis, I., Makridis, J. & Mastorakis, G. (2019) Location-Based Social Network Data for Tourism Destinations, in M. Sigala, R. Rahimi, and M. Thelwall (ed.), *Big Data and Innovation in Tourism, Travel, and Hospitality*, Singapore: Springer Nature, pp. 105-114.

Wise, N. & Heidari, H. (2019) Developing smart tourism destinations with the Internet of Things, in M. Sigala, R. Rahimi & M. Thelwall (ed.), *Big Data and Innovation in Tourism, Travel, and Hospitality*, Singapore: Springer Nature., pp. 21-29.

Yamamoto, M. (2019) Furthering big data utilization in tourism, in F. P. García Márquez and B. Lev (ed.), *Data Science and Digital Business*, Switzerland: Springer Interanational, pp. 157-171.

10 Customer relationship management through blockchain technology

Ugur Tandogan

Introduction

Mass production techniques and mass marketing increased the availability of products for customers and changed the competition environment in the middle of the 20th century. Businesses became less familiar with their customers in this period. In other words, customers were only numbers in the eyes of businesses that make mass production and sales. Extreme increase in the product and service options in the market has caused businesses not to deal with the customers' needs separately. There was no difference between such businesses in the eyes of customers. However, many businesses abandoned mass marketing concept in the following years and started to increase long-term customer loyalty and reestablish the connections with their current customers for this purpose. Some businesses activated their competition and increased success in the market by implementing relational marketing principles within the scope of the strategical and technological-based customer relationship management (CRM) (Chen & Popovich, 2003).

The fourth industrial revolution, which started as Henning Kafgermann and Bosch Company presented the study to the German Government in 2012 and is referred as the Industry 4.0 (Turan, 2020), requires the digital transformation of many business activities in a comprehensive and detailed way (Sayar, 2019). The targets of Industry 4.0 also include meeting the requests of personal customers. The transition from mass customer services to personal product and service production is facilitated by combining smart production and industry components (Pisching *et al.*, 2015). The increased communication possibilities from technological developments have removed the borders between countries and increased the globalization rate. Today's world, which can be named the digital interaction era, has revealed the blockchain technology in line with the requirements and conditions of the era (Sayar, 2019). It has been seen that customer relations managements have preferred to collect customers' data using blockchain systems recently.

The main reasons for this preference are that it enables storing data more safely, sharing data with relevant units and making analyses more easily (Turkiye Bilisim Vakfi, 2019a). Conveniences provided by the digital era, such as the transition to customer-based production and blockchain, have increased the importance and effectiveness of CRM activities for tourism businesses. Digitalization has greatly accelerated due to the pandemic period we have been living in as people mandatorily stayed in their houses and the majority of jobs were carried out remotely. However, the pandemic, which accelerated digitalization, affected the world's economy and tourism movements quite negatively. According to the United Nations World Tourism Organization (UNWTO) (2020a), international tourism movements decreased 82% in 2020 compared to the June-December period in the previous year. UNWTO (2020b) predicts that these developments will accelerate the relevant steps of tourism activities, that take pioneering steps in digitalization.

The transfer of activities to digital platforms provides opportunities that will increase CRM activity thanks to blockchain technology. Some of these are opportunities that will increase customer satisfaction and loyalty such as providing closeness and trust with customers, making smart contracts, rapid access to and safe sharing of information, reducing costs, safe storage and evaluation of customers' information, and payment ease and advantage with a token system. Current studies state that the increase in digitalization will not be limited to the pandemic period and will continue afterward. According to the Association of Turkish Travel Agencies (TURSAB, 2020), studies showed that 75% of people who used digital channels for the first time during the pandemic will continue to use these after the pandemic.

Literature review

The fact is that businesses focusing on preserving and gaining customer loyalty can increase their market share as they transform customer relations into a structure with different value dimensions. The main reason for a customer to continue purchasing a product and service from a business is the customer perceived value. Customer perceived value is the positive difference between the product benefit and the cost of obtaining it (Uzunoglu, 2007). Knowing the effects of customer value is of great importance for customer relations management, because the perception that the business values customers more than its competitors will attach customers to the business and increase the competitive power of the business (Ay, 2014). Graf and Maas (2008) state that one of the customer value approaches is the determination of the value by the customer. They also state that a key aspect of any product or service is the perception regarding how much it meets the goals and requests of the customer or how much it satisfies the customer. According to Rossanty *et al.* (2019), how much a customer is satisfied with a product is determined with the customer's expectations from the product and the benefit the product provides to the customer. If the product performance is greater than the expectation, the customer will be satisfied with the product or business; thus, the customer's loyalty tendency to the business will increase. Loyal customers will be less sensitive towards the pricing of the business and the business will have a positive image in the eyes of customers.

The strategic role of CRM will increase as customer-oriented businesses focus on customer-centered strategies instead of product-centered strategies (Onaran *et al.*, 2013). CRM is defined as a strategy that aims to develop and maintain customer relations (Rossanty, 2019). It is a strategy that ensures reducing the costs of an organization, increasing its profitability and enhancing customer loyalty (Saha & Sharma, 2020). After the developments in the customer-focused understanding, many marketing concepts have been started to be questioned. After CRM solutions, marketing experts have started to pay attention to not only gaining customers but also ensuring the loyalty of the current customers to the business (Demir & Kirdar, 2012). CRM is management efforts made to understand the requests and characteristics of customers and to improve relationships with customers by combining businesses processes and technology. The basis of CRM is the transition from product-focused marketing understanding to customer-focused marketing understanding.

Customer relations activities have affected and changed businesses and many customers while entering the millennium era (Kim & Suh, 2003). CRM is the entirety of applications, strategies, and technology used by the business to manage and analyze customer interactions and data during the customer lifecycle (Rouse, 2020). It is also a customer-centered relation management philosophy that enables coordination and cooperation between all units of the business, customers and business partners (Bozgeyik, 2005). According to Gordon (2002), it is a series of strategies and processes that create new and mutual values for customers and businesses, make sure the business is preferred, and focus on improving the services provided in their lifelong relations with the costumer. Nam * Lee (2019) states that CRM used to be based on conventional data analysis but it has had a transformation based on big data. According to Aktan (2018), technological development and digitalization enabled institutions and industries as well as individuals to produce great quantities of data rapidly with their smart devices. As a result, we have encountered a rapid increase in the diversity and volume of data that has never been seen in human history. This situation introduced the concept of big data.

In data-based customer analysis, the data obtained from customer interactions are used to get a better idea about customers' motivations, preferences, characteristics and behaviors. Thus, businesses are able to optimize their marketing channel performances, improve customer experiences and provide customer-specific products. It is possible to gain higher profits with more satisfied customers (Djteam, 2020). According to Fickel (1999), practices used in CRM technology connect customer touch points to both front office functions such as sales, marketing and customer relations, and back office functions such as finance, operations, logistics and human resources. Chen & Popovich (2003) states that the customer touch points of a company are the channels it connects with customers. For example, the Internet, e-mail, sales, direct mail, telemarketing operations, call centers, advertising, fax, pagers, stores and kiosks are touch points. These points are generally controlled by separate information systems.

Tourism businesses and customer relations management

It is necessary for hotels to find as many customers as possible and determine accurate strategies and targets to preserve these customers while expanding the market. Each hotel competes to provide the best service and to leave the best impression on customers. Positive or negative opinions of customers obtained as a result of CRM activities will shape the image of the hotel. If the hotel management and customers are in a compatible relation, positive opinions occur. CRM activities make sure to perpetuate the hotel guests' satisfaction and improve it (Rossanty, 2019). The competition environment of hotels causes positive developments in CRM activities. CRM combines three departments including marketing, sales and customer services in hotel businesses. However, it is seen that activities mainly focus on sales and this is generally actualized as money exchange. Additionally, hotels need to add operations, the fourth term, in the sales and customer services process. Operational activities are the fulfillment of special interests and needs of the customers before they come to the hotel, during their stay in the hotel and after they leave the hotel (Zunenshine, 2020).

CRM is also important for travel agencies. For instance, one encounters many service options from countless different firms to purchase a travel tour. Therefore, the travel preferences of a customer can move from a business to another very rapidly. It is important to know the requests of a customer to attract him/her to the business. CRM, thus, collects and evaluates the necessary data. Recording the individuals who contacted the business and purchased services, and their features enable the effective and efficient realization of CRM activities (Goksal, 2017). Similar situations are observed in restaurant businesses. A CRM system of a restaurant helps managing critical guest data that enables personalized service and marketing, effectively. Personalized services are tailored activities that make guests feel special and cause them to prefer the same business every time. It is extremely important for restaurant businesses, because, according to a study conducted by Thanx inc., 70% of the restaurant customers are disloyal (PR Newswire, 2015). Sharmen (2018) revealed the importance of CRM in restaurant businesses where CRM activities are generally not performed. Restaurants that perform effective CRM activities are preferred by customers. CRM is a part of an effective and efficient management system and is necessary for restaurants to keep their customers.

Digitalization in customer relationship management

Digitalization of CRM is the transfer of the customer data to a digital platform where these can be read and processed by the computer system and the conduct of the management on this platform. This platform where CRM will be transferred brings multidirectional advantages to both tourism businesses and customers.

Digital customer relationship management is the use of internet communication channels and technologies to improve customer relationships and enhance management activities. Digital CRM tries to determine the habits and preferences of customers as well as facilitating personalization and automated messaging (Rouse, 2020). The digital revolution is a good opportunity to promote loyalty among customers and optimize customer relationships. It can be possible to digitalize the conversations between the business and customers within the scope of CRM (Efficy, 2020). Quality standards in customer services are re-determined with the developing technology. Customers seek faster and problem-specific assistance in this period. Companies determine the size of their potential by accurately using the data obtained from customer behaviors, and get a few steps ahead of their competitors by producing target-specific products and services. Closely following artificial intelligence, machine learning and blockchain like technologies, which are among the trends of digital transformation process, and including them in CRM process are mandatory in recent years. Artificial intelligence and machine learning are the most important matters in the period of change in customer relations. Smart systems that generalize data and improve themselves by using these all the time, are the biggest help for companies in terms of CRM. In fact, companies that foresee the expectations of consumers by evaluating customer behaviors get ahead of their competitors with the effective services they provide. Virtual assistants build a bridge between customer and company as well as facilitating the operations for both sides. Artificial intelligence, which closes the gap with the employees day by day, has started to be the primary preference of customers because of the services it provides (Aynaci, 2019).

According to the study by Gartner Inc., 73% of CRM expenses in the world in 2018 were made for software services. Expenditure on CRM software increased by 15.6% in 2018 and reached 48.2 billion dollar. According to the study results, CRM continues to be the biggest and fastest growing enterprise application software category. The worldwide enterprise application software revenue increased 12.5% in 2018 compared to the previous year and rose to around 193.6 billion dollar. Revenue obtained from CRM software constitutes 25% of this total revenue (Gartner, 2019). Gartner also foresees that the customer service industry, which developed in the digital transformation process, will continue to experience major change until 2025. It is foreseen that artificial intelligence will become prominent in this change and will provide major added value and savings to companies. Other predictions are as follows: Two-third of customer problems will be solved by artificial intelligence instead of humans in 2023. The operating profitability of companies that include artificial intelligence in customer relations will increase by 25%. 10% of companies will use blockchain technologies. Smart bot applicants, which are currently used around 15%, will reach 70%. These developments and predictions show the future importance of artificial intelligence, blockchain and similar technological trends in the inevitable digital transformation period (Aynaci, 2019).

Automation applications used in successful CRM provide conveniences such as improving the customer interactions of travel agencies, hostel businesses, boutique hotels, all-inclusive hotel businesses, five-star hotels, and restaurant businesses in the tourism industry, increasing their customer interactions and accessing customers' data quickly. Applications used in CRM are solution-oriented functional tools especially for businesses operating in the tourism industry and focusing on providing services. Just like all other industries, it is important for the tourism industry to follow technological developments, manage customer relationships well, and maximize the profitability by saving time. The compliance of the tourism business administrations with constantly changing and developing dynamics in technology will also facilitate CRM applications (Globalnet, 2019).

Blockchain

Blockchain is a distributed database in a network formed by many participants (Ko & Verity, 2016). It is a platform that enables transactions to be recorded in one network and monitored through the shared accounting book (Andersen, 2016). Cryptocurrency made a name for itself with Bitcoin. Blockchain was invented by a person (or a group of people) using the name Satoshi Nakamoto to serve as a public trading book for cryptocurrency bitcoin in 2008 (Wikipedia, 2020). In the article published by Nakamato in 2008, he introduced a cryptographic evidence-based electronic payment system without a third party validating tool to provide trust between the parties. Thus, both parties will be able to make a direct transaction reciprocally securely (Nakamato, 2008). It was, then, understood that blockchain could be used in all operations requiring correct data recording and transfer. Blockchain allows rapid, trackable and safe transfer of cryptocurrencies, information and digital properties (Ko & Verity, 2016). The blockchain system has two important concepts called public and private key. An account containing cryptocurrency in blockchain is expressed a public key. The person who owns the corresponding private key is the owner of this account. (Anderson *et al.*, 2016). The public key concept is similar to the account number of a bank account. Knowing this account number is not sufficient to access the money in the account. In addition to the account number, the password of the account must also be known. The password is the private key. The person who owns the private key can access to the cash in the account. Onal (2020) stated in his study that the public and private keys in the blockchain are used not only for crypto money accounts, but also for accessing accounts where information, contracts or certificates are kept.

According to notification stated in World Economic Forum that blockchain is related to a value exchange, no third parties are needed to manage these processes, and the system aims to facilitate and record the exchange of money and other properties (World Economic Forum, 2018). From another perspective, blockchain is in a way a database. Data are recorded in blocks in order of time in here. Once a block is complete, the next block can be created and the blocks are chained to each other. When parties in the blockchain network perform their transactions with encrypted identities, each transaction is distributed to many users' computers in a decentralized network. Blockchain is called distributed ledger technology because the processing of records is not tied to a single center and transactions occur at multiple points. In this system, data can only be hacked if at least 51% of all computers connected to the network are seized (Guven & Sahinoz, 2018). However, being a large number of computers on the network and the difficulty of decrypting make it nearly impossible to seize 51% of computers and hack transactions on the blockchain. It has high accuracy and reliability as the data in a block is verified by all computers in the network and becomes unalterable. Pilkington (2017) states that the accuracy, transparency and reliability of the information is high, as CRM companies collect customers' comments about businesses through the blockchain system.

Blockchain in the digitalized tourism industry and its usage in CRM

The digitalized tourism industry uses platforms and facilities like e-commerce, information, online purchasing, marketing, website analysis, e-research methods (Calvaresi *et al.*, 2019), mobile hotel booking (Ozturk & Khaldoon, 2017) and RFID (Radio Frequency Identification) technologies (Ozturk & Hancer, 2014) and recently the blockchain. The rise of digital customers, the growth of artificial intelligence techniques, the heterogeneity of digital platforms create a new dimension for trust and security. Trust, accepted as the basic building block of society, plays an important role in the formation and strengthening of commercial interactions and relations. Unfortunately, existing mechanisms cannot deal with the formation of strategic lies, malicious behaviors and deceptive coalitions. Since a central asset or a mediator cannot solve these problems, an important technological innovation that will close this gap is needed. Blockchain technology which is considered the driving force of the technological revolution created important opportunities for companies in the tourism industry in this regard (Calvaresi *et al.*, 2019). The trustworthiness of blockchain will undoubtedly affect the tourism industry positively.

The distributed ledger system in the blockchain and the memorandum of understanding will ease the concerns about trust between the parties. Thus, the safety protocols of blockchain will reach all tourism stakeholders and create an environment convenient for brands in the industry to remain popular, and improve the tourism experience (Rejeb & Rejeb, 2019). There are different safety expectations of hotel management and customers during accommodation services. For example, hotel management will want the customer to be a compliant person, not cause economic damage to the hotel, not disturb other customers, not to involve in criminal incidents during the stay, and not to damage the hotel's image with his/her behaviors. On the other hand, the customer will want the information shared by the hotel that she encountered while choosing the hotel and expected to meet her expectations to be true, and the services she paid for this purpose to be provided (Morkunas & Paschen, 2019).

Three important application of blockchain for CRM in the tourism industry, digital identity, coin and token, and smart contracts are reviewed below.

Digital identity is the electronic equivalent of the identity card. It is an electronic certificate that is used to prove an individual's identity and right to access online information and services (Docusign, 2020). Previously, individuals used their appearance and the witness of their acquaintances to prove their identities. More formal tools were needed for national and international processes as governments institutionalized and the borders between the countreies were formed. Today, individuals have to prove who they are with physical documents such as an identity card, driving license, bills, etc. The need for introducing oneself has risen in many places. This situation has caused some difficulties such as the difficulty of sharing physical documents, loss of time, and the security gap due to sharing the information openly. Digital identities have begun to be used to obtain maximum benefit from the opportunities provided by digitalization in today's world. Individuals can share their identity information with different institutions on digital medium; thus, all processes like registration, and application can be actualized on a digital medium. The main processes where digital identity can be used in tourism industry are as follows: hotel-restaurant reservation and check-in, purchasing travel ticket and check-in, travel insurance, renting cars, drones, bicycles, etc., visa application processes, money transfer, purchasing tickets/loyalty card for events, age control in events, gambling games, consignment processes (e.g. interpretation headsets of a museum), etc.. Since blockchain is a safe database that can be accessed by all users on the network, it provides an important opportunity for users about sharing necessary identity information where needed. It is possible to keep a part of personal data on the blockchain and keep other parts on systems other than the blockchain and/or on the blockchain, and to keep data verification, abstraction, encrypted versions to prevent data to be disclosed without authorization (Turkiye Bilisim Vakfi, 2019b). Tourists will complete formality processes with various institutions safely and rapidly by using digital identity on blockchain, and will enjoy their travel more. Tourism businesses will obtain the opportunity to do their preparations and processes faster by closely following their customers through digital identity. Additionally, they will be able to manage customer-focused services by determining the preferences of customers they follow.

Coin and Token - Coin is a cryptocurrency, digital and virtual money that allows secure cryptographically encrypted transactions and was put on the market as an alternative to regular currencies (Garipagaoglu & Demir, 2020). Securing digital currency with cryptography makes the possibility of forgery and double-spend almost impossible (Frankenfield, 2020). Coin is a digital currency with its own blockchain. It has other money features without having a physical presence. You can purchase products and services, make payments and transactions with coin. The firstly-launched and most popular coin is Bitcoin. Later, coins like Ethereum, Litecoin and Ripple were launched. A *token* is a digital asset. However, it does not have its own blockchain,but uses the blockchain infrastructure of a coin. Moreover, it is not possible to make a payment or purchase products and services with token. It can be used for only a specific purpose within a certain structure. For a better understanding, token may be likened to a concert ticket. You can only go to a concert with this ticket and cannot make other types of payments (Sistemcoin, 2020).

Businesses in the tourism industry produce smart contracts and tokens that can only be used in their organization and facilitate their customers' foreign exchange swap processes, reduce the exchange costs, closely monitor their preferences and obtain the opportunity to get loyal customers by providing advantages to customers via token.

Smart Contracts and Smart Records - Contracts are written agreements containing the handover of certain assets and values from one person to another based on conditions between people and objects. A third party verifies the transaction in these contracts or arbitrates them. The contract-based control of the transactions in the smart contracts between two or more institutions is the verification of the computer in the blockchain network instead of a central mediator or arbitrator (Marabito, 2017). Smart contracts promise a precise and safe handover on blockchain. According to the study by Gartner firm, the transaction quality will increase 50% until 2023 via smart contracts (Jones, 2020). Smart contracts enable making transactions within a previously-limited flow on the data in the blockchain network. Parties agree on the content of the contract and the contract is uploaded to the blockchain network after cryptographically signed. Smart contracts are currently not legally binding (Turkiye Bilisim Vakfi, 2019c).

Smart records are smart contracts that are used for a more specific area and enable making changes in the content. For instance, a smart record in travel industry contains many data related to holidays and travels (contract articles, costs, class, commissions, included services, etc.). These smart records can be customized and changed more easily compared to conventional methods and contracts. For instance, an economy class flight ticket has a free baggage allowance of 15 kg. It might be necessary to change the flight class to increase this free baggage allowance to 30 kg. This process can be turn into a situation that can cause additional costs, time and workforce such as drawing a new ticket for a new flight class in the conventional system. However, the free baggage allowances can be easily changed via the smart records of the travel agency and any possible change in value can be calculated immediately. Similarly, transferring a ticket to another customer by changing the name on the ticket is not possible for airline businesses that operate in conventional system today.

However, such change and transfer will be made on blockchain via smart records. If allowed by the relevant protocol, smart records can be separated and used if wanted. For example, a passenger who bought a turnaround flight ticket for a destination will be able to turn this into two separate tickets and use these at different times if wanted. Similarly, a full pension hotel reservation can be turned into a half pension or bed-and-breakfast concept if wanted. Price changes specified in the contract conditions for such situations will be calculated automatically and reflected to the check. Smart records are canceled when the service is used or at the end of its life, provided that it is determined by the protocols created (Further Network, 2019). Smart records can calculate the price/value of many services related to passengers and tourists through a token defined in the system. When it will be possible to process smart records as determined and make the necessary payments within the tourism ecosystem with token, it will be possible to reduce transactions, reduce costs and consequently increase customer loyalty (Yildiz & Tanriverdi, 2019). Each copy of contracts made on the blockchain is distributed to every user in the network and its accuracy is ensured with a mathematical system. Therefore, human weaknesses such as error or cheating is not a matter in these processes and the contract is based on an extremely safe foundation for both parties.

Conclusion

Sharing information is of great importance in tourism activities. It is necessary to share information with many different organizations during a trip. For instance, the information of the travelling customer needs to be shared with the travel agency, airlines and other transportation firms, luggage carriage organization, security units, the accommodation place, etc. Problems that may occur while sharing this information will interrupt the customer's travel, and concern and unhappiness will replace the joy that would be felt during the travel. When this information is stored with the blockchain technology, all the users in the network will verify these items of information and the organizations who reach the network will access the safe information fast. In addition, customers will access the information on the blockchain, such as travel, transportation and accommodation while making a travel plan.

More importantly, the information shared on the network will be verified by all users and the security of the information will be high as it is visible to all. Tourism enterprises will be able to record the needs and preferences of the customers they serve in the blockchain system and use them in customer loyalty management. (Kripto Para Turkiye, 2019). A hotel or a travel agency will be able to follow the route the customers take from their home to the airport until they check-in for the flight and even arrive at the hotel via digital identity and transparent information sharing on the blockchain. This opportunity will enable firms to reduce the waiting duration for check-in and other problems; thus, customer satisfaction will be increased. At first glance, monitoring the customers' movements can be seen as an intervention to private life. However, the ability to arrange the level of shared information according to the customer's preferences in the blockchain eliminates this problem. Thus, blockchain technology will be able to fulfill customers' requests without interfering with their private lives without any problems (Dogru *et al.*, 2018).

Blockchain technology has a multi-directional operating system in tourism activities. For instance, the tourist experience will improve thanks to blockchain. The level of learning with technology use between tourism operators and tourists will increase. Tourist destinations and their expenditures can be monitored and recorded in the blockchain database as much as the identity owner allows as a result of digital identities. The relevant tourism businesses in the network will be able to determine the purchasing profiles and preferences of the tourists from this information, with the help of the transparency and accessibility of the information on the blockchain. Payments that will be made via blockchain network and cryptocurrencies will be away from banking limitation, easier, and transboundary transfers will be fast and low cost. Decreasing the number of mediators in the service provision thanks to blockchain will decrease the general operation costs of tourism businesses. This will be beneficial for both the businesses and customers (Kwok & Koh, 2019).

The use of blockchain with CRM software will provide an important level of data security. With the integration of CRM with blockchain, organizations will have verifiable records and the access of unwanted followers to the resources of the business will be blocked. Since data are stored in blocks in the blockchain technology, the personal information of a person, transaction details and other information can be collected in another block (Fourtoy, 2020). Thus, CRM can develop strategies based on customer features. Blockchain closes the important gap of the CRM infrastructure by providing accurate customer data cost-effectively. Additionally, it can improve customer service with the connections it establishes in different platforms. The solution of creating a real customer profile is not one-dimensional. Thus, it is necessary to constantly update all customer-related data collection sources to enter data, access them, analyze them easily, securely share them, and ensure cooperation between businesses and sectors. New infrastructure such as blockchain is needed to overcome the slow speed and prohibitive costs of conventional CRM technology and to adapt rapidly to changing market trends and customer needs (Jin, 2020).

One of the most important factors that motivate the use of blockchain technology in CRM is that both service purchasers and suppliers can benefit from blockchain-based systems. The elimination of the mediator and provision of income increase by saving costs are important advantages for the supplier. However, since the elimination of the mediator is noticed by costumers, they may not expect a decrease in prices. CRM should make customers notice the advantages obtained via the use of blockchain in their firms. If the fact that blockchain is a trust-based system, the information obtained from it is accurate and their information will be used for their benefit is noticed as an advantage by the customer, customer satisfaction and loyalty can be ensured (Calvaresi *et al.,* 2019).

Advantages brought by digitalization have made tourism services easier, more efficient and comprehensive, economic and more attractive. These developments have created smart travel possibilities, smart destinations and new profession waves (UNTWO, 2020b). There are some obstacles before the common use and mass adoption of blockchain in today's world. There are various elements that need to be digitalized before proceeding to blockchain. Additionally, all stakeholders need to accept and embrace this structure. Making decisions that will pave the way for the use of blockchain by government administrations, institutions or managers will speed up the process. Despite all the negative impacts of the pandemic, it had a positive impact in speeding up this process and raising awareness about the advantages of the blockchain (Turkiye Bilisim Vakfi , 2020).

Blockchain constitutes one of the most important technological developments in the digitalization process. Tourism businesses that include blockchain technology into their customer relations processes will be able to increase their efficiency. Similar to businesses, customers have started to prefer smart systems instead of humans. The most important reason of this preference is that smart systems do not have human weaknesses, in addition to conveniences they provide. Accessing safer information with lower possibility of error more easily and lower cost in short time is actualized with blockchain, which forms the infrastructure of smart systems. Customers who used to have to wait in line on the phone to get information and support about products, are now able to use a mobile application with the digitalization and can solve their problems with artificial intelligence without much waiting.

Owing to the advantages of digitalization and blockchain technology, tourism enterprises have had the opportunity to provide customer-based services of the new era. Customers will prefer these businesses that provide accurate information, deal with their requests closely, provide solutions for the problems that may occur rapidly by using blockchain technology and will be the loyal customers of these businesses. Businesses that focused on the innovations and opportunities provided by the digital era about customer relations will take important steps in increasing efficiency by transferring CRM activities to the blockchain system. One of the most important features that will help tourism industry to survive during and after pandemic is safety, since, people prefer to travel and stay in safe and healthy places. This features should be taken into consideration more carefully than ever before. Blockchain technology will make it possible to do this easily and safely. Health information of people and hygiene information of transportation channels, touristic facilities, and other tourism services will be presented to the access of all users on the blockchain, which is a distributed, transparent and safe system. CRM will be able to keep the information about their businesses, personnel, the destination of the business, and the hygiene measures in the transportation channels up-to-date by sharing them with the customers all the time. All this information will give confidence and motivation to potential customers.

Discussion questions

1. What are some obstacles to the widespread use and mass adoption of blockchain in tourism industry today?
2. What kind of infrastructure should tourism enterprises have in order to use the blockchain system in CRM?
3. Can a tourism business use blockchain technology to change the customer's perceived value?
4. Blockchain technology will provide a lot of convenience to tourism industry during and after the pandemic. Has the pandemic process also provided an advantage to blockchain technology?

References

Aktan, E. (2018) Buyuk veri: Uygulama alanları, analitigi ve guvenlik boyutu, *Bilgi Yonetimi*, **1** (1), 1-22.

Andersen, N. (2016) Blockchain technology a game-changer in accounting?, *Deloitte*.

Anderson, L., Holz, R., Ponomarev, A., Rimba, P. & Weber, I. (2016) New kids on the block: An analysis of modern blockchains, *arXiv prepint arXiv*, 1606.06530.

Ay, E. (2014) *Musteri iliskileri yonetimi araci olarak sadakat kartlarinin musteri memnuniyeti ve bagliligina etkileri uzerine bir arastirma*, tuksek lisans tezi, Selcuk Universitesi, Sosyal Bilimler Enstitusu, Konya.

Aynacı, T. (2019) *Dijital Donusum Musteri Iliskilerinde Standartları Yeniden Belirliyor*, Piariletisim, https://www.piariletisim.com/dijital-donusum-musteri-iliskilerinde-standartlari-yeniden-belirliyor [Accessed: 18 December 2020].

Bozgeyik, A. (2015) CRM Ne Icin Onemlidir?, https://abdullahbozgeyik.wordpress.com/2015/06/25/bayi-yonetimi-sitesi-yazim-crm-nicin-onemlidir/ Retrieved 10.11.2020.

Calvaresi, D. & Leis, M. (2019) Trust in tourism via blockchain technology: Results from a systematic review article in information, *Technology & Tourism.*

Chen, I. J., & Popovich, K. (2003). Understanding customer relationship management (CRM): People, process and technology. *Business Process Management Journal,* **9** (5), 672-688.

Demir, F. O. & Kirdar, Y. (2012) Musteri iliskileri yonetimi: CRM, *Review of Social, Economic & Business Studies,* **7** (8), 293-308.

Djteam, (2020) *What is data-driven marketing?*, DemandJump.com, https://www.demandjump.com/blog/what-is-data-driven-marketing, Accessed: 28 October 2020.

Docusign (2019) *What is digital id?*, docusign.com, https://support.docusign.com/en/articles/What-is-a-digital-ID [Accessed: 26 December 2020].

Dogru, T., Mody, M. & Leonardi, C. (2018) Blockchain technology and its implications for the hospitality industry, *Boston Hospitality Review,*1-12.

Efficy (2020) *CRM and Digital Transformation*, www.efficy.com/crm-and-digital-transformation-why-and-how-should-you-digitize-your-customer-relationship/ [Accessed: 12 October 2020].

Fickel, L. (1999) Know your customer, *CIO Magazine,* **12** (21), 62-72.

Fourtoy, C. (2020) *How does blockchain technology benefit customer relationship management?*, Blockchainara, https://blockchainara.com/how-does-blockchain-technology-benefit-customer-relationship-management/ [Accessed: 1 December 2020].

Frankenfield, J. (2020) *What is Cryptocurrency*, Investopedia, https://www.investopedia.com/terms/c/cryptocurrency.asp [Accessed: 2 January 2021].

Further Network (2019) *From passenger systems to systems for passengers, Transforming transactions in travel ecosystem with blockchain*, https://www.further.network [Accessed: 11 November 2020].

Garipagaoglu, U. N. & Demir, E. (2020) Crypto coins and blockchain: An evaluation on the tourism sector, *Journal of Yasar University*, **15** (58), 210-220.

Gartner (2019) *Gartner says worldwide customer experience and relationship management software market grew 15.6% in 2018*, Gartner Newsroom, https://www.gartner.com/en/newsroom/press-releases/2019-06-17-gartner-says-worldwide-customer-experience-and-relati [Accessed: 5 January 2021].

Globalnet (2019) Turizm Sektorunde CRM, https://blog.globalnet.com.tr/turizm-sektorunde-musteri-iliskileri-yonetimi/ [Accessed: 2 November 2020].

Gordon, I. (2002) Best practices: Customer relationship management, *Ivey Business Journal*, **67** (2), 1-5.

Goksal, S. (2017) *Turizmcilere tavsiye: 'CRM'de fark yaratan hizli yol alir*, https://www.turizmgunlugu.com/2017/12/26/turizmcilere-tavsiye-crmde-fark-yaratan-hizli-yol-alir/ [Accessed: 8 October 2020].

Graf, A. & Maas, P. (2008) Customer value from a customer perspective: A comprehensive review, *Journal für Betriebswirtschaft*, **58** (1), 1-20.

Guven,V. & Sahinoz, E. (2018) *Blockzincir Kripto Paralar Bitcoin*, Istanbul: Kronik Kitap.

Jin, F. (2020) *The Blockchain Approach to Customer Relationship Management*, https://cointelegraph.com/news/the-blockchain-approach-to-customer-relationship-management [Accessed: 1 December 2020].

Jones, L. C. (2020) *Gartner Predicts That Organizations Using Blockchain*, Gartner News Room, https://www.gartner.com/en/newsroom/press-releases/2020-01-30-gartner-predicts-that-organizations-using-blockchain- [Accessed 21 November 2020].

Kim J. & Suh E. (2003) A model for evaluating the effectiveness of CRM using the balanced scorecard, *Journal of Interactive Marketing*, **17** (2), 5-19.

Ko, V. & Verity, A. (2016) *Blockchain for the humanitarian sector: Future opportunities, dh network*, https://blockchain4aid.org/resources/blockchain-for-the-humanitarian-sector-future-opportunities/ [Accessed:16 November 2020].

Kripto Para Turkiye (2019) *Turizm Sektoru blockchain teknolojisinden nasil yararlanabilir*, https://www.kriptoparaturkiye.com/turizm-blockchain/.html [Accessed: 20 December 2020].

Kwok, A. O. J. & Koh, S. G. M. (2019) Is blockchain technology a watershed for tourism development?, *Current Issues in Tourism*, **22** (20), 2447-2452.

Morabito, V. (2017) *Business Innovation Through Blockchain*, Cham, Springer International Publishing AG.

Morkunas, V. J. & Paschen, J. (2019) How blockchain technologies impact your business model, *Business Horizons*, **62**, 295-306.

Nakamoto, S. (2008) Bitcoin: A peer-to-peer electronic cash system, https://bitcoin.org/bitcoin.pdf [Accessed:14 December 2020].

Nam, D. and Lee, j. (2018) Business Analytics Use İnpopovich CRM: A Nomological Net From IT Competence To CRM Performance, *International Journal of Information Managemen*t, **45** (1), 233-245. https://doi.org/10.1016/j.ijinfomgt.2018.01.005 19.11.2020.

Onal, H. (2020) SSL kullanımı ve sayısal sertifika, imza isleri, *Dokumanlar*, 1-16.

Onaran, B., Bulut, Z.A. & Ozmen, A. (2013) A study to investigate the effect of customer value on customer satisfaction, brand loyalty and customer relationship management performance, *Business and Economics Research Journal*, **4** (2), 37-53.

Ozturk, A. B. & Khaldoon, N. (2017) Understanding mobile hotel booking loyalty: An integration of privacy calculus theory and trust-risk framework, *Information System Frontie*r, **19** (4), 753–767.

Ozturk, A.B. & Hancer, M. (2014) Hotel and IT decision-maker characteristics and information technology adoption relationship in the hotel industry, *Journal of Hospitality and Tourism Technology*, **5** (2), 194-206.

Pilkington, M. (2017) Can blockchain technology help promote new tourism destinations? The example of medical tourism in Moldova, *SSRN Electronic Journal.*

Pisching, M.A. & Junqueira, F. (2015) Service composition in the cloud-based manufacturing focused on the industry 4.0., in *Technological Innovation for Cloud-Based Engineering Systems*, Springer, pp. 65-72.

PR Newswire (2015) *Study finds 70% of retail and restaurant customers never make a return visit*, https://www.prnewswire.com/news-releases/study-finds-70-of-retail-and-restaurant-customers-never-make-a-return-visit-300027410.html [Accessed: 18 December 2020].

Rejeb, A. & Rejeb, K. (2019) Blockchain technology in tourism: Applications and possibilities, *World Scientific News*, **137** (2019), 119-144.

Rossanty, Y., Nasution, M.D.T.P. & Irawan, I. (2019) Communication and customer relations strategy in improving hotel guests' satisfaction in Samosir tourism area, North Sumatra, Indonesia, *International Journal of Management*, **10** (6), 88–98.

Rouse, M. (2020) *CRM: Customer Relationship Management*, TechTarget, https://searchcustomerexperience.techtarget.com/definition/CRM-customer-relationship-management. [Accessed: 3 November 2020].

Saha, P. & Sharma, A. (2020) Creating value through customer relationship management in the automotive industry in India, *International Journal Of Research In Commerce & Management*, **11** (9), 1-13.

Sayar, S. (2019) *Dijitallesme Ile Yeni Olusan Kavramlar: Endustri 4.0, Iot Ve Blockchain Uygulamaları*, Yuksek Lisans Tezi, Maltepe Universitesi Sosyal Bilimler Enstitüsü, Istanbul.

Sharmen, S. (2018) Impact of customer relationship management practices on customer behavior in restaurant industries of Bangladesh, *International Journal of Marketing Research Innovation*, **2** (1), 47-63.

Sistemcoin (2020) *Coin ile Token Arasındaki Fark Nedir?*, https://blog.sistemkoin.com/coin-ile-token-arasindaki-fark-nedir/ [Accessed: 25 December 2020].

Turan, M. (2020) *Endustri 4.0 ne zaman basladi?*, www.bilisimhocasi.com/endustri-4-0-ne-zaman-basladi [Accessed: n.d.].

Turkiye Bilisim Vakfi (2019a) *Binance blockchain tabanli CRM agına yatırım yapti*, Blockchain Turkiye, https://bctr.org/binance-blockchain-tabanli-crm-agina-yatirim-yapti-9644/ [Accessed: 12 November 2020].

Turkiye Bilisim Vakfi (2019b) *Blokzinciri teknolojisi terminoloji calismasi*, Blockchain Turkiye, https://bctr.org/dokumanlar/Blokzinciri_Teknoloji_Terminoloji.pdf [Accessed: 28 October 2020].

Turkiye Bilisim Vakfi (2019c) *Digital kimlik raporu*, Blockchain Turkiye, https://bctr.org/dokumanlar/Dijital_Kimlik.pdf [Accessed: 29 October 2020].

Turkiye Bilisim Vakfi (2020) *Covid-19 sonrasında bizi bekleyenler ve blockchain'in yeni donemdeki yeri*, Blockchain Turkiye, https://bctr.org/wp-content/uploads/2020/06/Covid19-Blockchain-Makale-BCTR.pdf [Accessed: 12 November 2020].

TURSAB (2020) *Covid-19 sürecinde Turkiye ve dunya turizmi degerlendirmesi,* https://www.tursab.org.tr/duyurular/covid-19-surecinde-turkiye-ve-dunya-turizmi-degerlendirmesi [Accessed: 4 January 2021].

UNWTO (2020a) *International tourism and Covid-19,* https://www.unwto.org/international-tourism-and-covid-19 [Accessed:15 December 2020].

UNWTO (2020b) *Digital transformation,* https://www.unwto.org/digital-transformation [Accessed: 15 December 2020].

Uzunoglu, E. (2007) Musteri odaklı pazarlama anlayışına gore deger yaratma: Bir model olarak deger iletim sistemi, *Eskisehir Osmangazi Universitesi IIBF Dergisi,* **2** (1), 11-29.

Wikipedia (n202) Blockchain, https://en.wikipedia.org/wiki/Blockchain [Accessed:18 October 2020].

World Economic Forum (2018) *Blockchain beyond the hype a practical framework for business leaders,* http://www3.weforum.org/docs/48423_Whether_Blockchain_WP.pdf [Accessed: 20 November 2020].

Yildiz, S. and Tanriverdi, H. (2019) Turizm ve seyahat endüstrisinde blok zinciri uygulamalari: Further.network projesi ornegi, *Uluslararasi Turizm Kongresi Bildiriler Kitabi,* 867-873.

Zunenshine, M. (2020) *15 Best Hotel CRM,* https://crm.org/crmland/best-hotel-crm [Accessed: 18 November 2020].

11 Extended reality and gamification in destination marketing organizations: Current perspectives and future opportunities

Mark Baker and Arthur Huang

Introduction

Technology plays an increasingly important role in enhancing destination marketing organizations' competitiveness (Trunfio & Della Lucia, 2019; Castro *et al.*, 2017; Li *et al.*, 2017). The technology-driven marketing tools that can promote tourism services, address customers' real-time needs, and co-create value at destinations are expanding (Samara *et al.*, 2020; Buhalis, 2019). In particular, extended reality and gamification demonstrate the potential for transforming destination marketing organizations' business models and strategies. They are becoming widely recognized as effective tools for stimulating tourists' interactions with destinations (Cranmer *et al.*, 2020; Lacka, 2020). While existing studies have broadly discussed the potential of extended reality in hospitality and tourism (e.g., Beck *et al.*, 2019; Wei, 2019) and gamification in management (e.g., Wanick & Bui, 2019), literature related to how extended reality and gamification technologies should be adopted and applied by destination marketing organizations remains limited. There is a need to guide destination marketing organizations' digital marketing strategies and operational adaptations from a theoretical perspective (Lacka, 2020; Aebli, 2019; Xu *et al.*, 2017). This research aims to fill this gap by systematically reviewing literature about extended reality and gamification in destination marketing to uncover how such technologies are being applied and evaluated.

Literature review

DMOs serve the dual purpose of marketing and managing destinations to attract and satisfy visitors worldwide; however, DMOs face significant challenges regarding how their value is perceived (Reinhold *et al.*, 2019). Criticism regarding destination marketers' struggles to adapt to the continually transforming tourism market is not a new phenomenon (Sheehan *et al.*, 2016); their utility is approaching an inflection point where they must evolve to survive (Pike, 2016; Reinhold *et al.*, 2015). Since their inception, the responsibilities of DMOs have expanded (Pike & Page, 2014). DMOs must adapt to satisfy visitors, strive toward sustainable methods, and ensure destinations' long-term competitiveness (Pearce, 2016). As the technological tools and tourists change, so too must the strategies of destination marketers.

Marketers must understand the fundamental differences between the behaviors, attitudes, and demographic characteristics of different groups (Reisenwitz & Fowler, 2019) to successfully reach each cohort by utilizing the most effective methods (Hamed, 2017). For DMOs, it is suitable to define groups or cohorts by using generational parameters (Reisenwitz & Fowler, 2019). Previously, people belonging to the Baby Boomer generation (born 1946-1964) (Patterson & Pegg, 2009) and Generation X (born 1965-1979) (Goh & Lee, 2018) served as the primary targets for marketing efforts within the tourism industry (Reisenwitz & Fowler, 2019; Patterson & Pegg, 2009). However, the focus must now shift from the Baby Boomers and Generation X to Generation Y, known as the Millennials (born 1980-1994). Millennials are more technologically savvy than their predecessors, and they rely on digital information and technology for travel planning significantly more than Generation X (Reisenwitz & Fowler, 2019). Given that Generation Z or iGen (born 1995-2009) will succeed the Millennials, DMOs must attempt to remain on the cutting edge and apply state-of-the-art technologies now more than ever (Goh & Lee, 2018). DMOs must evolve by revamping their products, services, and models to meet the needs of the more technologically literate and dependent generations (Hamed, 2017). Innovative technologies present DMOs with numerous opportunities to engage with their target markets in unique ways (Li *et al.*, 2017). Destination marketers need to embrace the advancements in technology-based visualization techniques to succeed in the digital era (An *et al.*, 2021). Extended reality technologies particularly demonstrate significant promise for destination marketing (Lee *et al.*, 2020; Marasco *et al.*, 2018).

Extended Reality (XR) and Gamification

Extended reality (XR) refers to various types of immersive technologies that rely on virtual and physical environments, including mixed reality (MR), augmented reality (AR), augmented virtuality (AV), and virtual reality (VR) (Kwok & Koh, 2020; Ludlow, 2015). MR focuses on bridging the gap between reality and virtuality, where AR and AV are categorized (González *et al.*, 2021). AR projects computer-generated (virtual) content onto the real-world environment (Cranmer *et al.*, 2020; Yung & Khoo-Lattimore, 2019). For example, the AR game Pokémon Go projects Pokémon into users' real-world environments by using phones' cameras to capture the real-world environment and presents the digital augmentation via the phones' screen displays. AV is essentially the inverse of AR. AV projects real-world objects into virtual worlds (González *et al.*, 2021; Guo *et al.*, 2020), which can help train individuals while safely simulating stressful conditions, such as performing emergency maintenance tasks (Neges *et al.*, 2018). VR fully immerses users in virtual worlds that may simulate real-world environments or produce novel fantasy worlds (Araiza-Alba *et al.*, 2021; Makransky *et al.*, 2019). For instance, VR video games may involve users wearing head-mounted devices that fully cover their eyes. Their only view is that of a screen that presents them with an entirely virtual environment. Sound is supplied via headphones, which immerses users in experiences more fully. Additionally, users may interact with the virtual world via handheld controllers or stationary treadmills designed explicitly for user-controlled movements in virtual environments (e.g., Omni by Virtuix). XR technology has become more ubiquitous and viable (El-Jarn & Southern, 2020). It can be applied for various purposes, including training, safety, maintenance (Doolani *et al.*, 2020), education, and entertainment (Lalos *et al.*, 2020).

While XR can be utilized for practical purposes, XR experiences can also be gamified. Moreover, tourists' experiences can be gamified (Abou-Shouk & Soliman, 2021), even without any XR technology. Gamification is commonly defined as the inclusion of gaming elements in non-gaming contexts (Jahn *et al.*, 2021; Schöbel *et al.*, 2020). Deterding *et al.* (2011) propose the following five game design elements: interface (levels, leaderboards, badges); patterns/mechanics (limited time and resources, taking turns); principles/

heuristics (game styles, goals); models (curiosity, challenge, fantasy); and methods (playtesting, play-oriented). Gamification is not creating a full-fledged game; it only implies the inclusion of some game elements (Garcia *et al.*, 2019). The non-gaming context is essentially boundaryless; there is only one express exclusion, the utilization of game elements when creating a game (Deterding *et al.*, 2011); this allows almost any experience to be subject to some degree of gamification. Marketers are already using gamification techniques to manipulate consumers' behaviors and attitudes (Whittaker *et al.*, 2021). With smartphone penetration increasing globally, marketers have numerous opportunities to gamify experiences (Högberg *et al.*, 2019). In hospitality and tourism, research has found that gamifying destination experiences through smartphones generates positive outcomes for tourists and DMOs (Garcia *et al.*, 2019).

The literature has studied XR across various industries, such as healthcare (Sugimoto, 2021), architecture, engineering, construction (Alizadehsalehi *et al.*, 2020), design (El-Jarn & Southern, 2020), and hospitality and tourism (Kwok & Koh, 2020). The effects of gamification have also been investigated in education (Legaki *et al.*, 2020), psychology (Zainuddin *et al.*, 2020), healthcare (Floryan *et al.*, 2020), e-commerce (Xi & Hamari, 2020), and hospitality and tourism (Worimegbe *et al.*, 2020). While there is an increasing consensus on XR and gamification's potential for industry-specific applications, DMOs require new knowledge to effectively employ XR (Marasco *et al.*, 2018; Tussyadiah, 2015) gamification (Garcia *et al.*, 2019) strategies. DMOs are in a critical era of development. They must modify and reinvent their products, services, and business models to reach the more technologically literate and dependent generations (Hamed, 2017). There is a need for research-based intervention to help DMOs identify pathways to adopt new strategies and enhance operations (Lacka, 2020; Aebli, 2019), and the effects of XR and gamification on marketing practices require further examination (Abou-Shouk & Soliman, 2021; Lin *et al.*, 2020). A comprehensive review that identifies theories and methods for DMOs will support them in integrating XR and gamification technologies into business strategies. Against the established background, this study aims to support and further the goals of DMO practitioners and theorists by exploring the following research objectives:

1. iIdentify how XR and gamification are evaluated in the destination marketing context, and
2. Identify XR and gamification research opportunities for destination marketing scholars and practitioners.

Methodology

This study employs a multi-stage literature review methodology. The method comprises four main phases: (1) search, (2) filter, (3) screen, and (4) analyze. These phases and their respective procedures are executed sequentially to ensure a systematic and replicable process.

(1) **Search phase.** During this phase, the authors consider various research databases to include for the study. EBSCOhost's Hospitality & Tourism Complete and Elsevier's ScienceDirect databases are selected. Following the database selection, the authors develop and refine combinations of keywords in Boolean phrases to collect and identify literature related to destination marketing organizations, extended reality, augmented reality, augmented virtuality, mixed reality, virtual reality, and gamification. For this study's purposes, the same search string is applied to both databases, as the structure and parameters of the string are accepted in Hospitality & Tourism Complete and ScienceDirect. The string is defined as: 'destination marketing' and ('augmented reality' or 'augmented virtuality' or 'extended reality' or 'mixed reality' or 'virtual reality' OR 'gamification').

(2) **Filter phase**. This step involves applying the filter-level inclusion and exclusion criteria. The inclusion filter criteria limit this study's results to research published between 2017 – 2021 in academic journals with the full text available in English. At this level, the exclusion criteria are the inverse of the inclusion criteria.

(3) **Screening phase.** First, duplicate results are removed. Second, the authors perform a manual screening to examine studies for relevance. Each article's title, keywords, abstract, introduction, and methodology are evaluated during the screening process. The inclusion criteria capture studies that focus on marketing applications that involve XR or gamification for tourist destinations. The exclusion criteria focus on eliminating styles of papers such as literature reviews, research notes, perspectives, and commentaries. Lastly, studies that discuss XR or gamification outside of the destination marketing context are eliminated. The authors screen 122 articles, and 94 are excluded for failing to meet the inclusion criteria; thus, 28 articles are included for the final literature analysis.

(4) **Analysis phase.** This involves the authors examining the remaining 28 articles through a full-text analysis. The final phase evaluation provides increased sensitivity beyond titles, abstracts, keywords, introductions, and methodologies. The authors examine the findings, conclusions, practical implications, theoretical contributions, and future directions of each study

to derive research areas and recommendations related to XR and gamification in destination marketing. Figure 11.1 overviews the literature review process.

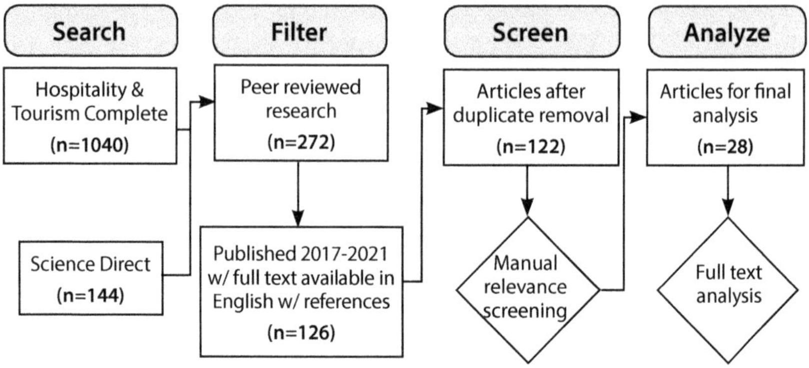

Figure 11.1: The four stages of the literature review process

Results

The number of studies identified per year is as follows: 2 (2017), 4 (2018), 7 (2019), 9 (2020), and 6 (status as of writing). There is an upward trend in the number of studies published regarding XR or gamification in the destination marketing context, excluding the year 2021. However, given that the number of published papers that regard this study's focus is six for 2021 at the time of writing, there is a significant opportunity for the number of published papers to grow throughout the remainder of 2021. Of the 28 studies included, three consider gamification aspects, and only one regards gamification in conjunction with any type of XR technology (AR). The remaining 25 studies focus on VR. Overall, the number and themes of studies can be represented in the following way:

1 (gamification and AR) + 2 (gamification) + 25 (VR) = 28.

Table 11.1 summarizes the articles and the adopted theories or concepts on applying XR and gamification to destination marketing.

Table 11.1: Summary of destination marketing literature focusing on XR or gamification

Journal	Topic	Research Focus
2017		
Journal of Destination Marketing & Management	Gamification AR	Multi-stakeholder involvement management
Journal of Travel & Tourism Marketing	VR	Attention interest desire action model, pleasure, arousal
2018		
Journal of Destination Marketing & Management	VR	Perceived visual appeal, emotional involvement, behavioral intention
Information Technology & Tourism	VR	Peceived realness, perceived vividness, behavioral intention
Leisure Studies	VR	Technology acceptance model, enjoyment, perceived control, attitude
Tourism Management	VR	Presence, attitude
2019		
Annals of Tourism Research	Gamification	Achievement motivation theory
Tourism Management	VR	Mental imagery, sense of presence, tourism brand experience
Journal of Travel & Tourism Marketing	VR	Stimulus organism response theory, technological embodiment, immersion, sensory stimulation, engagement, behavioral intentions
International Journal of Information Management	VR	Hedonic motivation system adoption model, flow theory, perceived easiness, perceived enjoyment
Journal of Destination Marketing & Management	VR	Technology acceptance model, motivation theory
e-Review of Tourism Research	VR	User-centered design
e-Review of Tourism Research	VR	Immersion, engagement, presence, sensory fidelity, emotion
2020		
Journal of Hospitality & Tourism Management	VR	Design play experience framework
Journal of Tourism Research	VR	Flow theory, arousal, control, attention, time distortion, enjoyment
Journal of Travel Research	VR	Stimulus organism response theory, authentic experience, attachment, affective response
International Journal of Hospitality Management	VR	Information system success model, vividness, attitude, telepresence, behavioral intention
Tourism Management	VR	Slow travel intention, nostalgia, destination image
Information Technology & Tourism	VR	Transportation imagery model, presence, attitude, purchase intention

Tourism Analysis	VR	Attitude, personal innovation, performance expectation
Tourism Recreation Research	VR	Technology acceptance model, perceived enjoyment, perceived immersion
Journal of Destination Marketing & Management	VR	Value-based adoption model, perceived benefits, perceived sacrifices, perceived value
2021		
Journal of Destination Marketing & Management	Gamification	Unified theory acceptance and use of technology, performance expectancy, effort expectancy, brand awareness, brand loyalty
Journal of Destination Marketing & Management	VR	Flow theory, stimulus organism response theory, telepresence, focused attention, temporal distortion
Journal of Hospitality & Tourism Management	VR	Media richness theory, vividness, interactivity, information sharing, information seeking
Tourism Management	VR	Attention interest desire action model, perceived usefulness, perceived credibility
Journal of Hospitality & Tourism Management	VR	Presence, emotion, engagement, behavioral intention
Tourism Recreation Research	VR	Perceived risk, destination image

Theories on supporting the use of XR and gamification in destination marketing

The existing theories on evaluating gamification and XR include the Unified Theory of Acceptance and Use of Technology (UTAUT) (e.g., Abou-Shouk & Soliman, 2021) and Achievement Motivation Theory (AMT) (e.g., Aebli, 2019). The UTAUT relies on four primary determinants: *performance expectancy* (perceived benefits resulting from technological adoption), *effort expectancy* (effort that must be exerted to use innovative technology), *social influence* (how the views or opinions of others influence one's perception of using new technology), and *facilitating factors* (level of support and type of resources available to support implementation); these components are used to predict adoption intention and usage for technological innovations (Abou-Shouk & Soliman, 2021; Venkatesh *et al.*, 2003). The AMT focuses on *achievement* (motive to do something better for oneself or in competition with others), *power* (need to influence others), and *affiliation* (desire to be with others) (McClelland, 1987). They are used to uncover social considerations that may arise when individuals interact with environments. Gamification is being explored via mobile phone applications that include game elements such as leaderboards, badges, points, and performance tracking, enabling vacation tourists to fulfill their needs for achievement, power, and affiliation (Aebli, 2019). The findings from the studies reviewed suggest that gamification can improve marketing firms' performance by positively impacting levels of engagement and behavioral intentions (Abou-Shouk & Soliman, 2021) while increasing motivation levels and satisfying tourists' social needs (Aebli, 2019).

These results indicate that there are theory-grounded benefits associated with adopting gamification practices for tourists' destination experiences. While gamification is being explored using the UTAUT and AMT perspectives, the number of studies and standpoints from which the concept is being explored within destination marketing must be expanded.

VR is being evaluated from various theoretical standpoints. For example, some studies (e.g., Vishwakarma *et al.*, 2020) are utilizing the Technology Acceptance Model (TAM) proposed by Davis (1985, 1989). The model revolves around the two central concepts of *perceived usefulness*, defined as *"the degree to which a person believes that using a particular system would enhance his or her job performance"*, and *perceived ease* of use, defined as *"the degree to which a person believes that using a particular system would be free of effort"* (Davis, 1989: 320). Studies using TAM find that perceived ease of use and perceived usefulness positively influence tourists' travel intentions and enjoyment (Li & Chen, 2019) and significantly impact individuals' intentions to use VR (Vishwakarma *et al.*, 2020). Other studies use Flow Theory to consider VR's destination marketing effects. Flow Theory revolves around achieving a state of equilibrium between challenges and skills where negative emotions are not perceived (Csikszentmihalyi, 1975); in other words, flow occurs when individuals are so fully immersed in an experience all worries melt away (Csikszentmihalyi, 1992).

Studies conducting investigations through the lens of Flow Theory (e.g., Huang *et al.*, 2020) find that tourists that participate in VR sightseeing experiences may experience flow states. Furthermore, VR-induced flow experiences positively affect satisfaction and visit intentions through the states of telepresence and focused attention (An *et al.*, 2021). Telepresence can be defined as the *"extent to which consumers feel their existence in the virtual space"* (Shih, 1998), and focused attention may be described as the narrowing of an individuals' focus to one particular activity or experience (Webster *et al.*, 1993).

To examine VR as a communication method, some studies (e.g., Lee *et al.*, 2021) use the Media Richness Theory proposed by (Daft & Lengel, 1986). Media richness determines how effectively and thoroughly a firm may communicate information through certain channels based on factors such as focus, communication ability, and language (Maity *et al.*, 2018; McGrath & Hollingshead, 1994); high levels of richness increase clarity and understanding (Maity *et al.*, 2018). Lee *et al.* (2021) find that VR can enhance media richness and positively impact consumers' information searches and sharing behaviors. This study agrees with Huang *et al.* (2013) that VR is an impactful and effective tool for destination marketers. However, a substantial amount of work must be completed to uncover additional benefits and address related challenges.

Future opportunities for destination marketing scholars and practitioners

A variety of research opportunities are derived based on the literature review. This study hopes to facilitate the advancement of gamification and XR agendas for destination marketing scholars and practitioners. First, VR must be explored from the physiological perspective more thoroughly (Huang *et al.*, 2020). For example, the effects of users' heart rate variations during VR experiences may assist research in uncovering how virtual experiences affect individuals' emotions when they interact with virtual campaigns (Marchiori *et al.*, 2018). The physiological standpoint may help discover why individuals' psychological perceptions and physiological reactions can contradict each other during VR experiences (Huang *et al.*, 2020). Users without any prior knowledge or experience with VR also require more attention (Vishwakarma *et al.*, 2020) because individuals possess varying levels of technological readiness and skill, which may affect how fully they may interact with or appreciate virtual experiences (An *et al.*, 2021; Lee *et al.*, 2021; Lee *et al.*, 2020; tom Dieck *et al.*, 2018). Furthermore, users may already be familiar with a destination to some extent, and how their level of familiarity impacts a virtual interaction with the destination must be more thoroughly understood (An *et al.*, 2021; Lee *et al.*, 2020; Rejón-Guardia *et al.*, 2020). Researchers must also expand the scope of social demographics that are being explored in the XR research (Yung *et al.*, 2021a; Huang *et al.*, 2020; Lo & Cheng, 2020; Lin *et al.*, 2020) to assess a broader range of people from different countries, cultures, age groups, and consumer groups (Abou-Shouk and Soliman, 2021; Weng *et al.*, 2021; Kim *et al.*, 2020).

Second, the various types of XR equipment and hardware available for commercial use require further analysis due to their varying levels of price, sophistication, and familiarity. For example, there are relatively simple and cost-effective ways to employ VR, such as Google Cardboard, and more advanced and expensive methods, such as Facebook's Oculus Rift and its variations (Yung *et al.*, 2021b). Outside of wearable headsets and goggles, other hardware, such as desktop computers and smartphones, can provide XR experiences, and the differences between the benefits and downsides of each should be compared (Lee *et al.*, 2020). In terms of the physical hardware features, factors such as the weight of the device, the screen size, and the ergonomics of the fit may play a role in tourists' perceptions and impact their virtual experiences, and the research must identify the impacts of such variables (Flavián *et al.*, 2019). In addition to examining consumers' experiences relative to the different types of equipment that can be used, the different types of XR experiences available to destination marketers must also be compared, such as VR and AR (Lee *et al.*, 2021; Kim & Hall, 2019; Tussyadiah *et al.*, 2018). For example, in some cases, AR may afford conveniences to consumers which VR does not, and vice versa (Evangelista *et al.*, 2020). Additionally, the duration of VR experiences must be examined to identify what the appropriate lengths of experiences are for different situations and visitors; this may help tourists avoid potential adverse side effects such as motion sickness or cybersickness if they remain fully immersed for too long (Li & Chen, 2019; Yung *et al.*, 2019). Scholars and practitioners must further consider XR's effects across the various stages of trips, such as pre-experience, experience, and post-experience, to determine when and how to maximize opportunities and value (Flavián *et al.*, 2019).

Third, existing theories on using gamification in destination marketing need to be expanded. Research that focuses on how gamification can increase the level of attachment tourists feel toward destinations is needed (Coghlan & Carter, 2020). Studies that utilize gamification to fulfill tourists' social needs can be executed to examine engagement and value co-creation in destinations (Aebli, 2019). While gamification certainly provides a new marketing and engagement avenue for DMOs, future research must also address the challenges when XR and gamification converge. Challenges related to collaboration, internal resources, funding, and development may arise while developing gamified AR experiences, such as an AR scavenger hunt (Lagiewski & Kesgin, 2017).

Fourth, a greater level of collaboration is required between the XR technology developers and destination marketers. Participatory design for maximizing value co-creation requires greater scrutiny (Marasco & Balbi, 2019). There is a lack of effective management frameworks for implementing contemporary destination experiences, and researchers and practitioners must focus on developing appropriate management structures (Lagiewski & Kesgin, 2017). To overcome such obstacles, Lagiewski and Kesgin (2017) describe that the multi-stakeholder involvement management (MSIM) framework, proposed by Waligo *et al.* (2013), may be appropriate to utilize when managing large destination projects. The MSIM focuses on bringing stakeholders together through three phases of attraction, integration, and management, which aim to form relationships and identify objectives among stakeholders to enhance the collaboration processes throughout the development of gamified destination experiences (Waligo *et al.*, 2013). In future work, stakeholders must investigate the feasibility of implementing more comprehensive managerial and collaborative frameworks to understand the operational impacts.

Conclusion

This study systematically reviews the literature regarding XR and gamification within the destination marketing context. DMOs face the challenge of pursuing new strategies to adopt innovative methods effectively. The authors believe that XR and gamification are complementary concepts that provide DMOs with new opportunities to engage their target markets. The amount of research regarding the state-of-the-art XR technology for DMOs is relatively small; however, this study indicates that XR holds significant promise for the future. As tourists become more familiar and comfortable with the concept of XR and the hardware required for the immersive experiences, DMOs must be prepared to engage tourists from around the world in new ways. At the time of this writing, destination research focuses on VR more heavily than other types of XR technologies. Still, other forms of XR may stand to provide significant benefits for destination marketers. The findings indicate that significantly more studies are needed to uncover the full potential of XR and gamification for DMOs. Future research may consider mixed methods, such as interviewing stakeholders and conducting meta-analyses to examine the theoretical and practical implications of XR and gamification for destination marketing.

Discussion questions

1. How can destination marketing organizations leverage XR and gamification to communicate with their target markets more effectively?
2. How may XR and gamification complement one another in destination marketing?
3. What features of XR and gamification should destination marketers consider when implementing the concepts into their marketing practices? How may a destination incorporate game design elements to increase its competitiveness? How would you go about selecting the most relevant game design elements to include?
4. What suggestions would you provide to enhance collaboration and communication among stakeholders seeking to implement XR or gamification at their destination? What challenges may arise while developing and integrating such experiences?
5. What experiences, if any, have you had with XR or gamification? What benefits and challenges did you perceive?
6. What form of XR do you believe would be the most effective for destination marketing? Why? What criteria would you use to evaluate XR and gamification for destination marketing?
7. How may a destinations' distinct characteristics (e.g., national park, world heritage site, or business center) shape how XR or gamification is incorporated into the destination marketing strategy?
8. What future trends do you anticipate relative to XR and gamification in destination marketing?

References

Abou-Shouk, M. & Soliman, M. (2021) The impact of gamification adoption intention on brand awareness and loyalty in tourism: The mediating effect of customer engagement, *Journal of Destination Marketing & Management*, **20**, 100559.

Aebli, A. (2019) Tourists' motives for gamified technology use, *Annals of Tourism Research*, **78**, 102753.

Alizadehsalehi, S., Hadavi, A. & Huang, J.C. (2020) From BIM to extended reality in AEC industry, *Automation in Construction*, **116**, 103254.

An, S., Choi, Y. & Lee, C.K. (2021) Virtual travel experience and destination marketing: effects of sense and information quality on flow and visit intention, *Journal of Destination Marketing & Management*, **19**, 100492.

Araiza-Alba, P., Keane, T., Chen, W.S. & Kaufman, J. (2021) Immersive virtual reality as a tool to learn problem-solving skills, *Computers & Education*, 164, 104121.

Beck, J., Rainoldi, M. & Egger, R. (2019) Virtual reality in tourism: A state-of-the-art review, *Tourism Review*, **74**, 586612.

Buhalis, D. (2019) Technology in tourism-from information communication technologies to eTourism and smart tourism towards ambient intelligence tourism: A perspective article, *Tourism Review*, **32**, 125138.

Castro, S., Silva, S.C. & Duarte, P. (2017) Does digital marketing really boost city tourism? Evidences from Porto's experience, *European Journal of Applied Business and Management*, **3** (3), 84-100.

Coghlan, A. & Carter, L. (2020) Serious games as interpretive tools in complex natural tourist attractions, *Journal of Hospitality and Tourism Management*, **42**, 258-265.

Cranmer, E.E., tom Dieck, M.C. & Fountoulaki, P. (2020) Exploring the value of augmented reality for tourism, *Tourism Management Perspectives*, **35**, 100672.

Csikszentmihalyi, M. (1975) *Beyond Boredom and Anxiety: The experience of play in work and games.* San Francisco, CA: Jossey-Bars Publishers.

Csikszentmihalyi, M. (1992) *The Psychology of Happiness,* London: Rider.

Daft, R.L. & Lengel, R.H. (1986) Organizational information requirements, media richness and structural design, *Management science*, **32** (5), 554-571.

Davis, F.D. (1985) *A technology acceptance model for empirically testing new end-user information systems: Theory and results*, doctoral dissertation, Massachusetts Institute of Technology.

Davis, F.D. (1989) Perceived usefulness, perceived ease of use, and user acceptance of information technology, *MIS Quarterly*, **13** (3), 319-340.

Deterding, S., Dixon, D., Khaled, R. & Nacke, L. (2011) From game design elements to gamefulness: Defining gamification, In *Proceedings of the 15th International Academic MindTrek Conference: Envisioning future media environments*, 9-15.

Doolani, S., Wessels, C., Kanal, V., Sevastopoulos, C., Jaiswal, A., Nambiappan, H., & Makedon, F. (2020) A review of extended reality (XR) technologies for manufacturing training, *Technologies*, **8** (4), 77.

El-Jarn, H. & Southern, G. (2020) Can co-creation in extended reality technologies facilitate the design process?, *Journal of Work-Applied Management*, **12** (2), 191-205.

Evangelista, A., Ardito, L., Boccaccio, A., Fiorentino, M., Petruzzelli, A.M. & Uva, A.E. (2020) Unveiling the technological trends of augmented reality: A patent analysis, *Computers in Industry*, **118**, 103221.

Flavián, C., Ibáñez-Sánchez, S. & Orús, C. (2019) Integrating virtual reality devices into the body: Effects of technological embodiment on customer engagement and behavioral intentions toward the destination, *Journal of Travel & Tourism Marketing*, **36** (7), 847-863.

Floryan, M., Chow, P.I., Schueller, S.M. & Ritterband, L.M. (2020) The model of gamification principles for digital health interventions: Evaluation of validity and potential utility, *Journal of Medical Internet Research*, **22** (6), e16506.

Garcia, A., Linaza, M.T., Gutierrez, A. & Garcia, E. (2019) Gamified mobile experiences: Smart technologies for tourism destinations, *Tourism Review*, **74** (1), 30-49.

Goh, E. & Lee, C. (2018) A workforce to be reckoned with: The emerging pivotal Generation Z hospitality workforce, *International Journal of Hospitality Management*, **73**, 20-28.

González, C., Solanes, J.E., Muñoz, A., Gracia, L., Girbés-Juan, V.&and Tornero, J. (2021) 'dvanced teleoperation and control system for industrial robots based on augmented virtuality and haptic feedback, *Journal of Manufacturing Systems*, **59**, 283-298.

Guo, Z., Zhou, D., Zhou, Q., Zhang, X., Geng, J., Zeng, S., Lv, C. & Hao, A. (2020) Applications of virtual reality in maintenance during the industrial product lifecycle: A systematic review, *Journal of Manufacturing Systems*, **56**, 525-538.

Hamed, H.M. (2017) Marketing destinations to millennials: Examining the compatibility between the destination marketing organization website and the millennial tourist prospects, *Journal of Tourism and Recreation*, **3** (1), 1-20.

Högberg, J., Shams, P. & Wästlund, E. (2019) Gamified in-store mobile marketing: The mixed effect of gamified point-of-purchase advertising, *Journal of Retailing and Consumer Services*, **50**, 298-304.

Huang, Y. C., Backman, S. J., Backman, K. F. & Moore, D. (2013) Exploring user acceptance of 3D virtual worlds in travel and tourism marketing, *Tourism Management*, **36**, 490-501.

Huang, X.T., Wei, Z.D. & Leung, X.Y. (2020) What you feel may not be what you experience: A psychophysiological study on flow in VR travel experiences, *Asia Pacific Journal of Tourism Research*, **25** (7), 736-747.

Jahn, K., Kordyaka, B., Machulska, A., Eiler, T. J., Gruenewald, A., Klucken, T., Brueck, R., Gethmann, C.F. & Niehaves, B. (2021) Individualized gamification elements: The impact of avatar and feedback design on reuse intention, *Computers in Human Behavior*, **119**, 106702.

Kim, M.J. & Hall, C.M. (2019) A hedonic motivation model in virtual reality tourism: Comparing visitors and non-visitors, *International Journal of Information Management*, **46**, 236-249.

Kim, M.J., Lee, C.K. & Jung, T. (2020) Exploring consumer behavior in virtual reality tourism using an extended stimulus-organism-response model, *Journal of Travel Research*, **59** (1), 69-89.

Kwok, A.O. & Koh, S.G. (2020) COVID-19 and Extended Reality (XR), *Current Issues in Tourism*, **23** (3), 1-6.

Lacka, E. (2020) Assessing the impact of full-fledged location-based augmented reality games on tourism destination visits, *Current Issues in Tourism*, **23** (3), 345-357.

Lagiewski, R. & Kesgin, M. (2017) Designing and implementing digital visitor experiences in New York State: The case of the Finger Lakes interactive play (FLIP) project, *Journal of Destination Marketing & Management*, **6** (2), 118-126.

Lalos, S., Kiourt, C., Kalles, D. & Kalogeras, A. (2020) Personalized interactive edutainment in extended reality (XR) laboratories, *ERCIM News Educ. Technol*, **120**, 29-30.

Lee, M., Lee, S.A., Jeong, M. & Oh, H. (2020) Quality of virtual reality and its impacts on behavioral intention, *International Journal of Hospitality Management*, **90**, 102595.

Lee, S. A., Lee, M. & Jeong, M. (2021) The role of virtual reality on information sharing and seeking behaviors, *Journal of Hospitality and Tourism Management*, **46**, 215-223.

Legaki, N.Z., Xi, N., Hamari, J., Karpouzis, K. & Assimakopoulos, V. (2020) The effect of challenge-based gamification on learning: An experiment in the context of statistics education, *International Journal of Human-Computer Studies*, **144**, 102496.

Li, T. & Chen, Y. (2019) Will virtual reality be a double-edged sword? Exploring the moderation effects of the expected enjoyment of a destination on travel intention, *Journal of Destination Marketing & Management*, **12**, 15-26.

Li, S.C., Robinson, P. & Oriade, A. (2017) Destination marketing: The use of technology since the millennium, *Journal of Destination Marketing & Management*, **6** (2), 95-102.

Lin, L.P. L., Huang, S.C.L. & Ho, Y.C. (2020) Could virtual reality effectively market slow travel in a heritage destination?, *Tourism Management*, **78**, 104027.

Lo, W.H. & Cheng, K.L.B. (2020) Does virtual reality attract visitors? The mediating effect of presence on consumer response in virtual reality tourism advertising, *Information Technology & Tourism*, **22** (4), 537-562.

Ludlow, B. (2015) Virtual reality: Emerging applications and future directions, *Rural Special Education Quarterly*, **34** (3), 3-10.

Maity, M., Dass, M., & Kumar, P. (2018) The impact of media richness on consumer information search and choice, *Journal of Business Research*, **87**, 36-45.

Makransky, G., Terkildsen, T.S. & Mayer, R.E. (2019) Adding immersive virtual reality to a science lab simulation causes more presence but less learning, *Learning and Instruction*, **60**, 225-236.

Marasco, A. & Balbi, B. (2019) Designing accessible experiences for heritage visitors through virtual reality, *e-Review of Tourism Research*, **17** (3), 426-443.

Marasco, A., Buonincontri, P., van Niekerk, M., Orlowski, M., & Okumus, F. (2018) Exploring the role of next-generation virtual technologies in destination marketing, *Journal of Destination Marketing & Management*, **9**, 138-148.

Marchiori, E., Niforatos, E., & Preto, L. (2018) Analysis of users' heart rate data and self-reported perceptions to understand effective virtual reality characteristics, *Information Technology & Tourism*, **18** (1), 133-155.

McClelland, D.C. (1987) *Human Motivation*, Cambridge, NY: Cambridge University Press.

McGrath, J.E. & Hollingshead, A.B. (1994) *Groups Interacting with Technology: Ideas, Evidence, Issues, and an Agenda*, Sage Publications, Inc.

Neges, M., Adwernat, S. & Abramovici, M. (2018) Augmented virtuality for maintenance training simulation under various stress conditions, *Procedia Manufacturing*, **19**, 171-178.

Patterson, I. & Pegg, S. (2009) Marketing the leisure experience to baby boomers and older tourists, *Journal of Hospitality Marketing & Management*, **18** (2-3), 254-272.

Pearce, D.G. (2016) Destination management: Plans and practitioners' perspectives in New Zealand, *Tourism Planning & Development*, **13** (1), 52-71.

Pike, S. (2016) Destination marketing organizations – Research opportunities in an era of uncertainty, In *International Conference on Tourism*, 21.

Pike, S. & Page, S.J. (2014) Destination marketing organizations and destination marketing: A narrative analysis of the literature, *Tourism management*, **41**, 202-227.

Reinhold, S., Beritelli, P. & Grünig, R. (2019) A business model typology for destination management organizations, *Tourism Review*, **74** (6), 1135-1152.

Reinhold, S., Laesser, C. & Beritelli, P. (2015) 2014 St. Gallen consensus on destination management, *Journal of Destination Marketing & Management*, **4** (2), 137-142.

Reisenwitz, T.H. & Fowler, J.G. (2019) Information sources and the tourism decision-making process: An examination of Generation X and Generation Y consumers, *Global Business Review*, **20** (6), 1372-1392.

Rejón-Guardia, F., García-Sastre, M.A., Orfila-Sintes, F. & Garau-Vadell, J.B. (2020) Virtual reality in tourism: Centennials acceptance, *Tourism Analysis*, **25** (2-3), 335-344.

Samara, D., Magnisalis, I. & Peristeras, V. (2020) Artificial intelligence and big data in tourism: A systematic literature review, *Journal of Hospitality and Tourism Technology*, **21**, 47-58

Schöbel, S., Janson, A., Jahn, K., Kordyaka, B., Turetken, O., Djafarova, N., Saqr, M., Wu, D., Söllner, M., Adam, M., Gad, P., Wesseloh, H. & Leimeister, J. M. (2020) A research agenda for the why, what, and how of gamification designs: Outcomes of an ECIS 2019 panel, *Communications of the Association for Information Systems*, **46** (1), 30.

Sheehan, L.R., Vargas-Sánchez, A., Presenza, A. & Abbate, T. (2016) The use of intelligence in tourism destination management: An emerging role for DMOs, *International Journal of Tourism Research*, **18** (6), 549-557.

Shih, C.F.E. (1998) Conceptualizing consumer experiences in cyberspace, *European Journal of Marketing*, **32** (7-8).

Sugimoto, M. (2021) Extended reality (XR: VR/AR/MR), 3D printing, holography, AI, radiomics, and online VR tele-medicine for precision surgery, *Surgery and Operating Room Innovation*, 65-70.

tom Dieck, D., tom Dieck, M.C., Jung, T. & Moorhouse, N. (2018) Tourists' virtual reality adoption: An exploratory study from Lake District National Park, *Leisure Studies*, **37** (4), 371-383.

Trunfio, M. & Della Lucia, M. (2019) Engaging destination stakeholders in the digital era: The best practice of Italian regional DMOs, *Journal of Hospitality & Tourism Research*, **43** (3), 349-373.

Tussyadiah, I.P. (2015)'Personal technology and tourism experience, In Egger, R. & Maurer, C. (Eds.), *ISCONTOUR 2015 Tourism Research Perspectives*, 1 - 10.

Tussyadiah, I.P., Wang, D., Jung, T. H. & tom Dieck, M. C. (2018) Virtual reality, presence, and attitude change: Empirical evidence from tourism, *Tourism Management*, **66**, 140-154.

Venkatesh, V., Morris, M.G., Davis, G.B. & Davis, F.D. (2003) User acceptance of information technology: Toward a unified view, *MIS Quarterly*, **27** (3), 425-478.

Vishwakarma, P., Mukherjee, S. & Datta, B. (2020) Antecedents of adoption of virtual reality in experiencing destination: A study on the Indian consumers, *Tourism Recreation Research*, **45** (1), 42-56.

Waligo, V.M., Clarke, J. & Hawkins, R. (2013) Implementing sustainable tourism: A multi-stakeholder involvement management framework, *Tourism Management*, **36**, 342-353.

Wanick, V. & Bui, H. (2019) Gamification in management: A systematic review and research directions, *International Journal of Serious Games*, **6** (2), 57-74.

Webster, J., Trevino, L.K. & Ryan, L. (1993) The dimensionality and correlates of flow in human-computer interactions, *Computers in Human Behavior*, **9** (4), 411-426.

Wei, W. (2019) Research progress on virtual reality (VR) and augmented reality (AR) in tourism and hospitality, *Journal of Hospitality and Tourism Technology*, **10** (4), 539-570.

Weng, L., Huang, Z. & Bao, J. (2021) A model of tourism advertising effects, *Tourism Management*, **85**, 104278.

Whittaker, L., Mulcahy, R. & Russell-Bennett, R. (2021) Go with the flow for gamification and sustainability marketing, *International Journal of Information Management*, **61**, 102305.

Worimegbe, P.M., Worimegbe, T.M. & Abiola-Oke, E. (2020) Gamification and customers' experience in the hospitality industry, *Journal of Tourism and Services*, **11** (21), 71-87.

Xi, N. & Hamari, J. (2020) Does gamification affect brand engagement and equity? A study in online brand communities, *Journal of Business Research*, **109**, 449-460.

Xu, F., Buhalis, D. & Weber, J. (2017) Serious games and the gamification of tourism, *Tourism Management*, **60**, 244–256.

Yung, R. & Khoo-Lattimore, C. (2019) New realities: A systematic literature review on virtual reality and augmented reality in tourism research, *Current Issues in Tourism*, **22** (17), 2056-2081.

Yung, R., Khoo-Lattimore, C. & Potter, L.E. (2019) VR the world: Investigating the effectiveness of virtual reality for destination marketing through presence, emotion, and intention, *e-Review of Tourism Research*, **17** (3), 368-384.

Yung, R., Khoo-Lattimore, C. & Potter, L.E. (2021a) VR the world: Experimenting with emotion and presence for tourism marketing, *Journal of Hospitality and Tourism Management*, **46**, 160-171.

Yung, R., Khoo-Lattimore, C., Prayag, G. & Surovaya, E. (2021b) Around the world in less than a day: Virtual reality, destination image and perceived destination choice risk in family tourism, *Tourism Recreation Research*, **46** (1), 3-18.

Zainuddin, Z., Chu, S.K.W., Shujahat, M. & Perera, C.J. (2020) The impact of gamification on learning and instruction: A systematic review of empirical evidence, *Educational Research Review*, **30**, 100326.

Index

www.ingramcontent.com/pod-product-compliance
Ingram Content Group UK Ltd.
Pitfield, Milton Keynes, MK11 3LW, UK
UKHW021828100926
8556IPUK00003B/226

9 781911 635475